Bio-Art™
to accompany

Biology

FOURTH EDITION

Solomon · Berg · Martin · Villee

Saunders College Publishing

Harcourt Brace College Publishers

Fort Worth   Philadelphia   San Diego   New York   Orlando   Austin
San Antonio   Toronto   Montreal   London   Sydney   Tokyo

A Note to the Instructor

Thank you for adopting *Biology*, Fourth Edition by Solomon, Berg, Martin, and Villee. We appreciate your use of our text and as a special aid for students we have produced Bio-Art™.

Bio-Art™ is a collection of important pieces of art from *Biology*, Fourth Edition rendered in black and white. Generally most pieces of Bio-Art™ do not include labels, so students can use the art as a learning tool. It is an excellent tool in measuring student understanding of processes and organisms.

This valuable study aid is free with copyright privileges to instructors who adopt *Biology*, Fourth Edition for class or can be purchased by students at a low cost. Please contact your bookstore if you would like Bio-Art™ to be purchased by students as a required or recommended supplement.

Suggested uses of Bio-Art™, for instructors who adopt *Biology*, Fourth Edition, include:

√ Copying Bio-Art™ as handouts for students, enabling students to label parts of figures, take notes, and avoid redrawing complicated diagrams, while the instructor uses a color overhead transparency from *Biology*, Fourth Edition

√ Duplicating Bio-Art™ on overhead acetates and referring to and writing on the acetates during lectures, while students refer to their own handouts

√ Distributing copies of Bio-Art™ as part of exams and quizzes

√ Having students complete homework assignments using their own copies of Bio-Art™

BIO ART™ LISTING

4-11	Plant cell
4-12	Animal cell
5-2	Plasma membrane
5-18	Glucose transport
8-6	Chloroplast
9-4	Mitosis/Cell cycle
9-12	Meiosis I & II
12-10	Initiation of translation
12-11	Translation - elongation cycle
13-1	Lac operon
13-2	Tryp operon
14-3	Recombinant DNA
14-5	Construction of genomic library
16-6	Development states of Drosophila from the egg to the adult fly
16-7	Diagram of imaginal discs and adult structures
17-11	Homologous organs
19-10	Gradualism versus punctuated equilibrium
21-8	Human and gorilla skulls
23-8	Structure of bacterial cell wall
24-6b	Paramecium
24-10b	Euglena
24-12	Life cycle of chalmydomonas
24-17a	Life cycle of water mold
25-2a	Life cycle of black bread mold
25-6a	Mushroom morphology
26-5	Moss plant
26-9	Life cycle of fern
27-3	Life cycle of pine
27-8	Structure of flower
28-2	Body plans/coelom
28-8	Structure of the Hydra
28-12	The common planarian
28-19	Nematode body plan
29-5	Clam anatomy
29-8	Internal structure of an earthworm
29-16	Grasshopper structure
30-14	Internal anatomy of bony fish
32-8	Variation in guard cells
33-4	Development and secondary growth of the stem
35-13	Pear flower and fruit
36-10	Darwin's experiments
37-5	Organ systems
38-6	Human skeleton
38-14	Human muscles, anterior view
38-15	Human muscles, posterior view
39-2	Structure of a multipolar neuron
40-6	Comparison of the brains of six vertebrate classes
40-7	Human brain
40-8	The spinal cord

40-9	The withdrawal reflex
40-11	Midsagittal section through the human brain
40-1	Dual innervation of the heart and stomach
41-8	Anatomy of the human ear
41-15	Structure of the human eye
42-6	Types of blood vessels
42-8	Evolution of the vertebrate heart
42-10	Section through human heart
42-16	Pattern of blood flow
42-17	Circulation of blood through some of the principal arteries and veins of the body
44-1	Types of respiratory structures found in animals
44-6	Human respiratory system
45-3	Earthworm digestive system
45-4	Human digestive system
45-10	Kidney structure
47-5	Activation of genes by steroid hormones
47-6	Second messenger mechanism
47-11	Principal endocrine glands
47-12	Hypothalamus/anterior lobe of the pituitary
47-15	Regulation of thyroid hormone secretion
47-18	Regulation of glucose concentration
48-3	Male reproductive system
48-4	Testis structure
48-6	Spermatogenesis
48-10	Female reproductive system
48-11	Female reproductive system -- anterior view
48-16	The menstrual cycle
49-12	Extraembryonic membranes
54-1	Carbon cycle
55-8	Greenhouse effect
55-9	UV radiation and the ozone layer

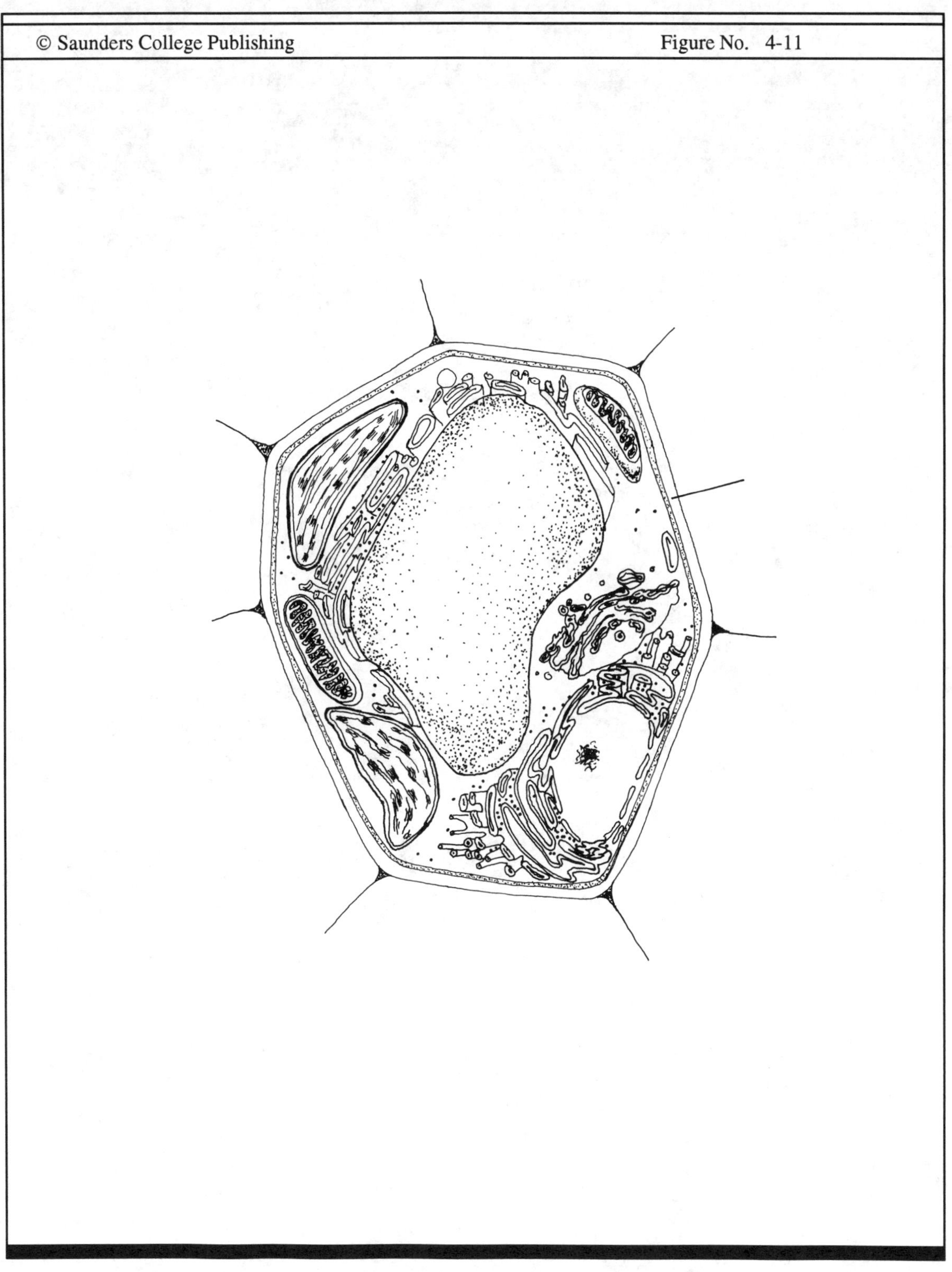

Figure No. 4-12

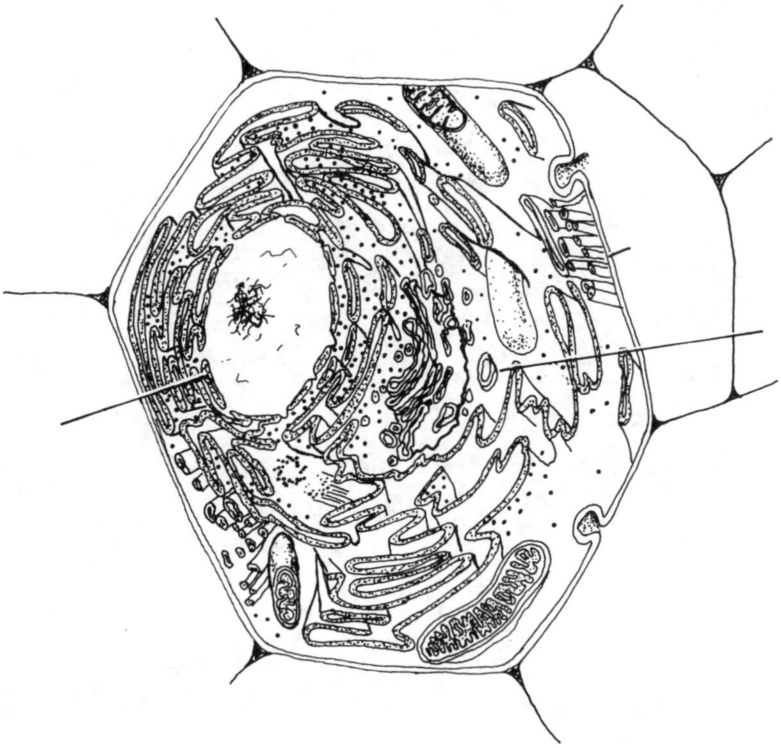

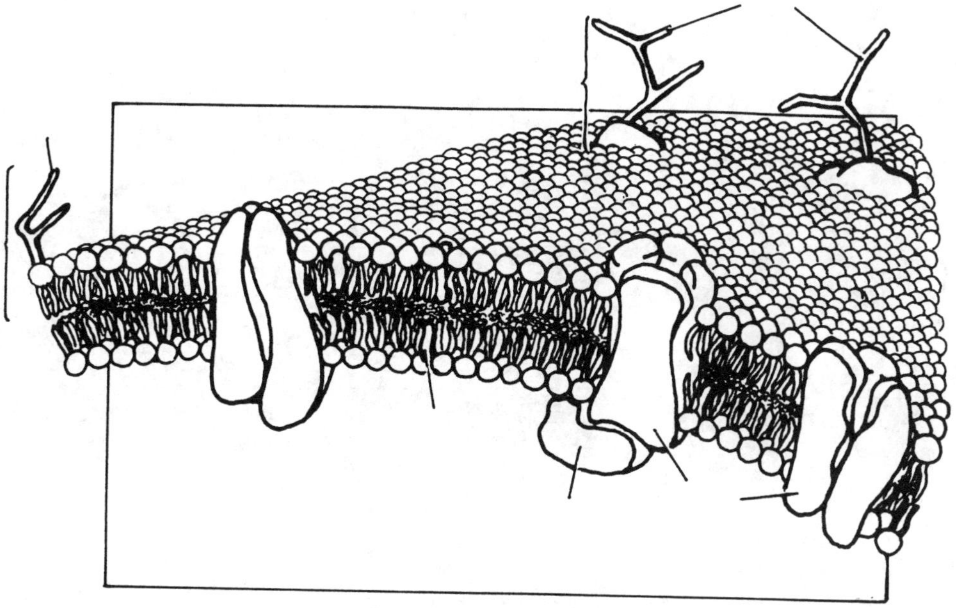

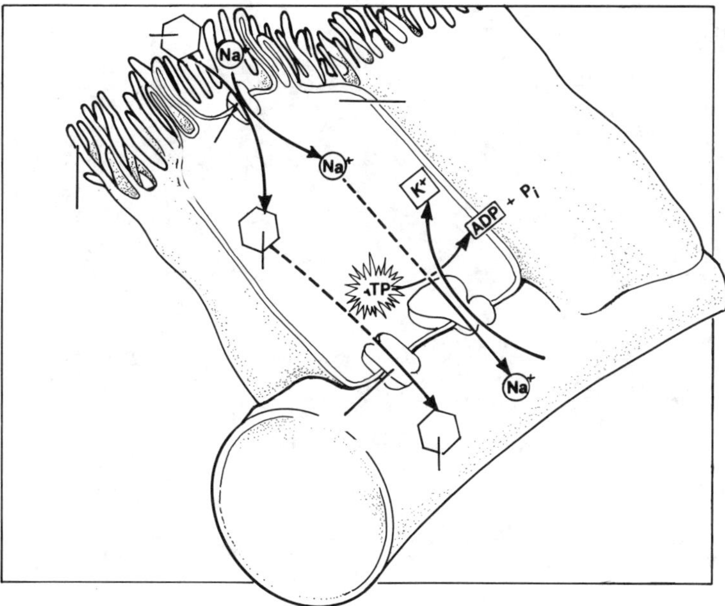

Na⁺
Na⁺
Na⁺
K⁺
ADP + Pi
TP

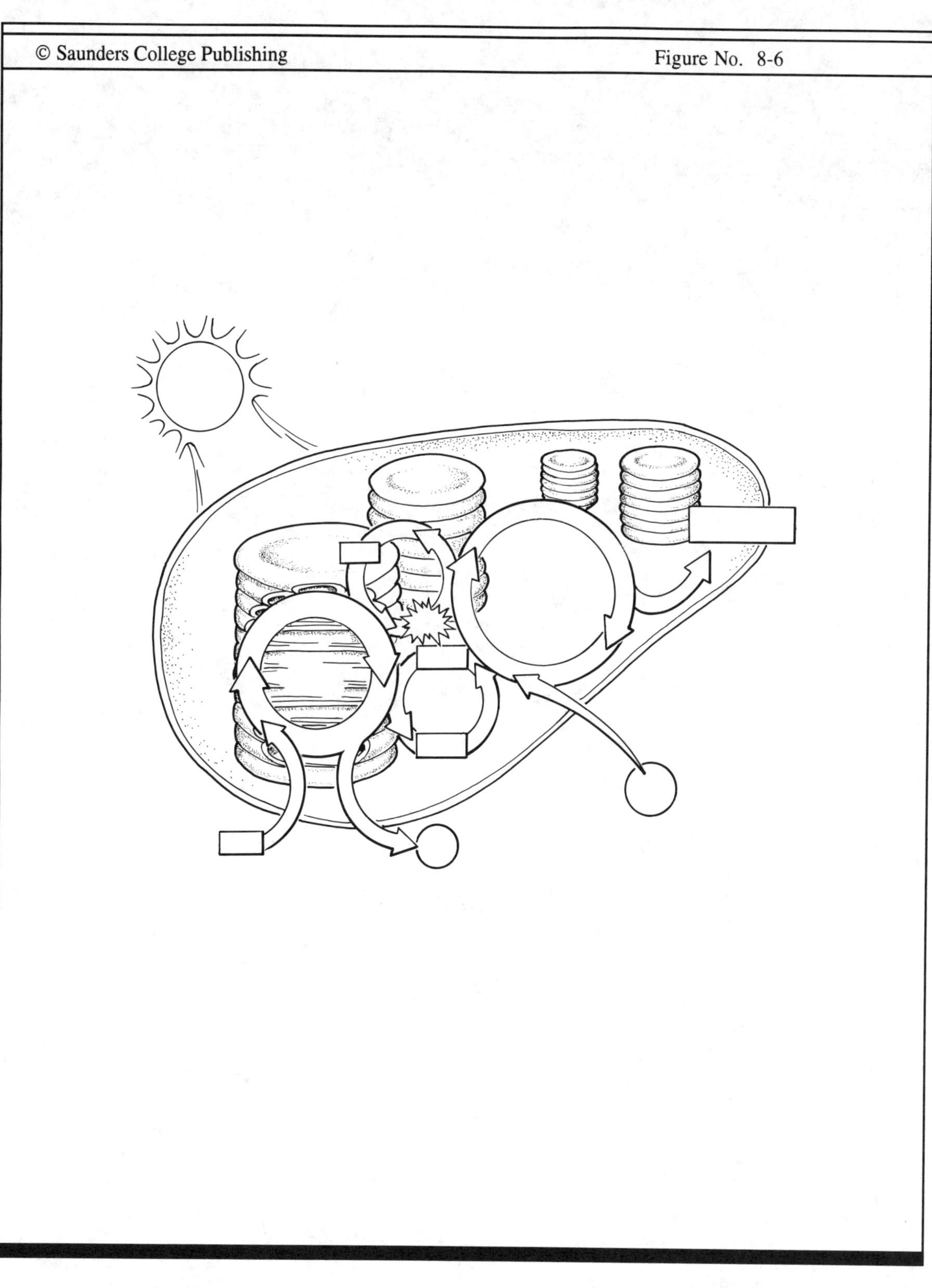

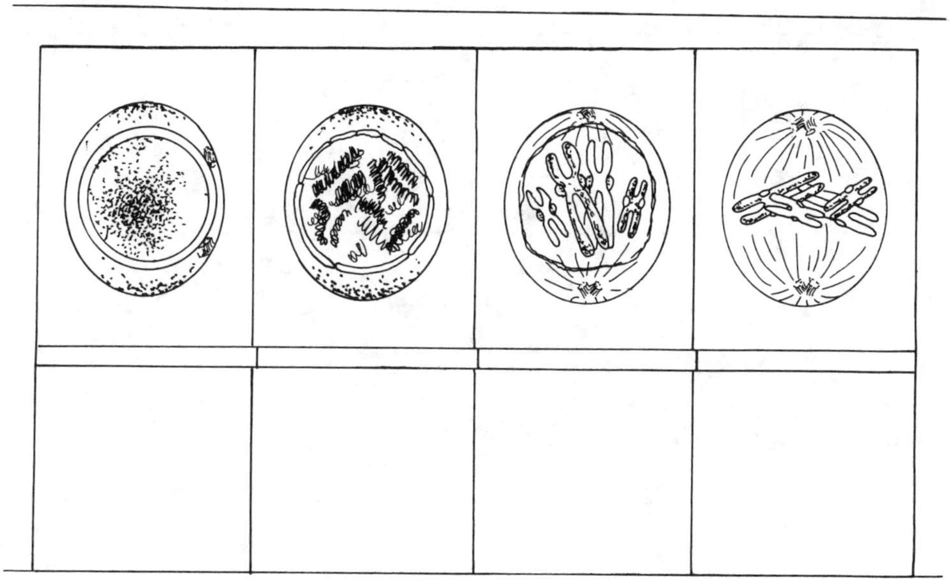

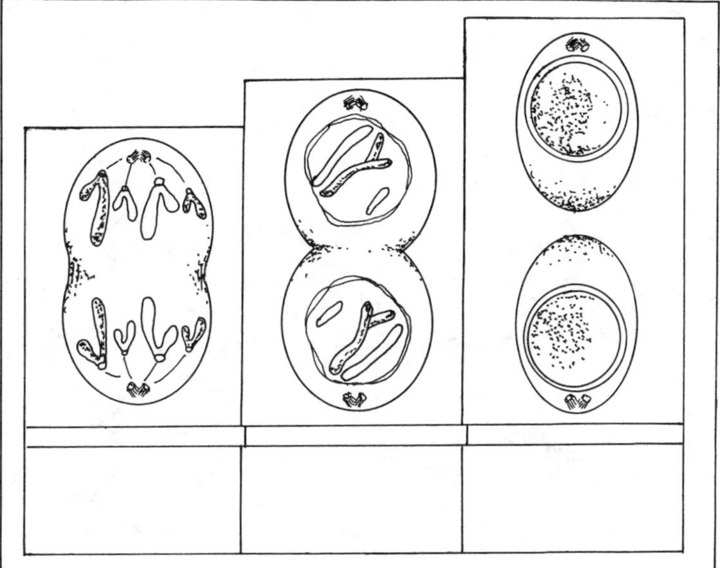

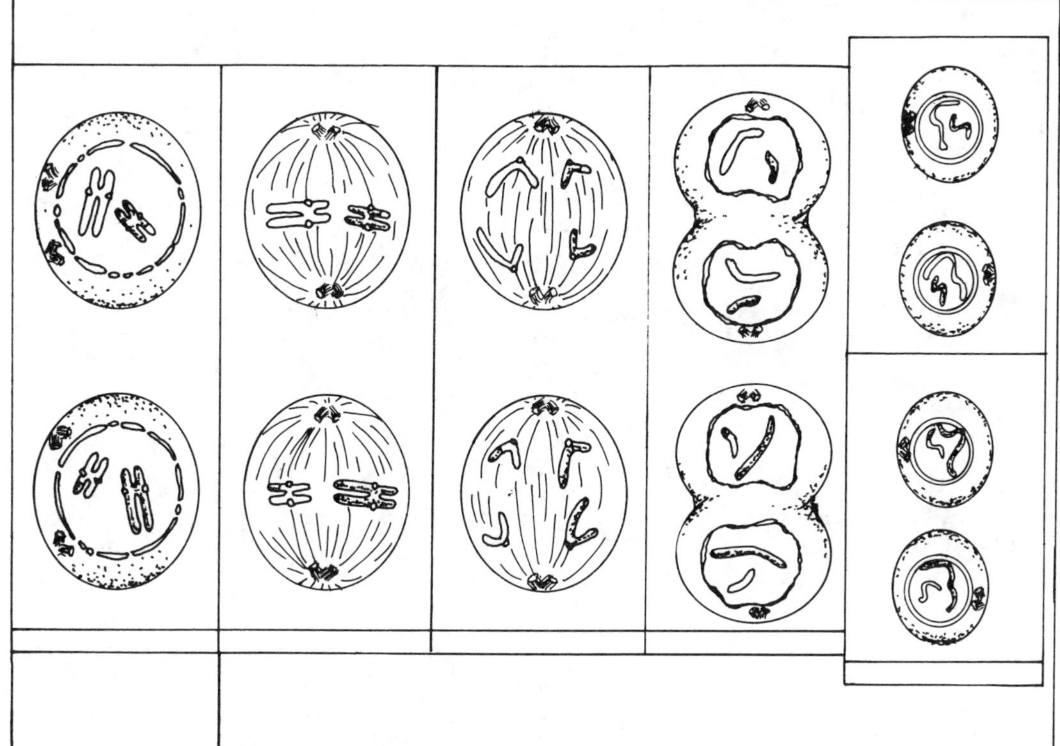

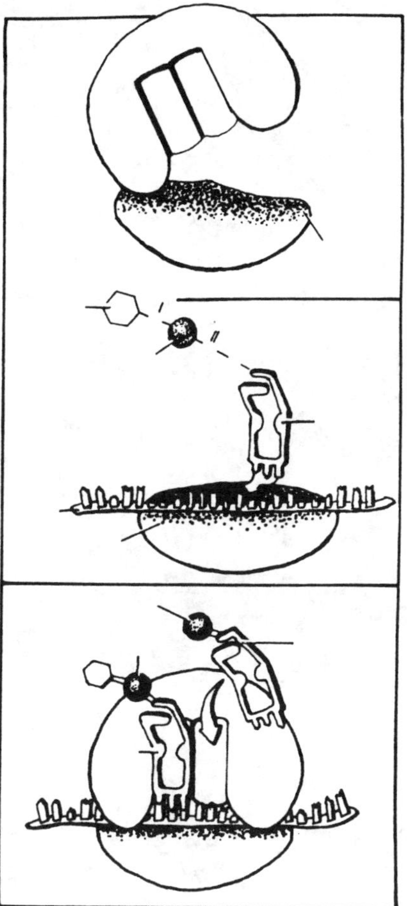

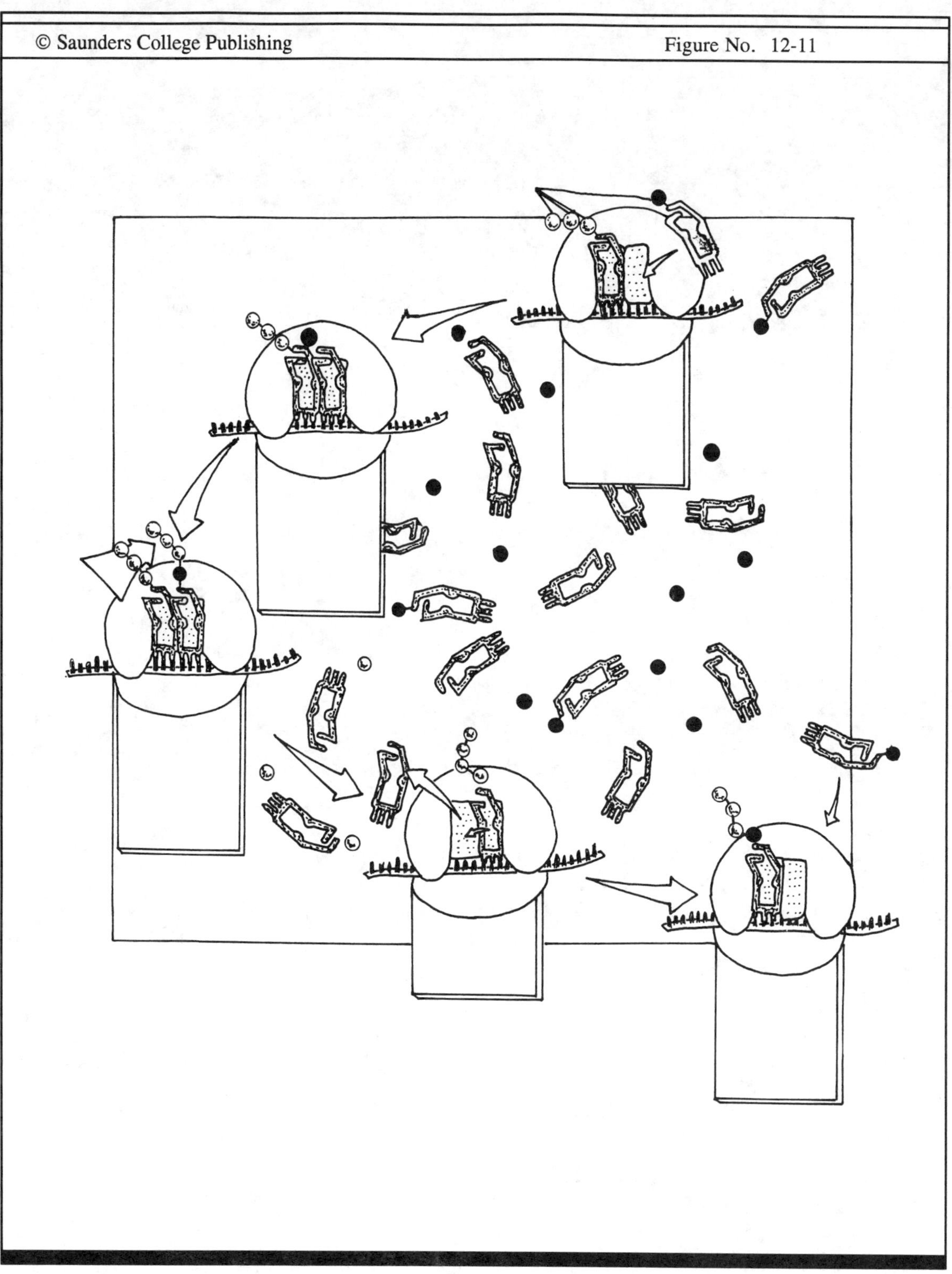

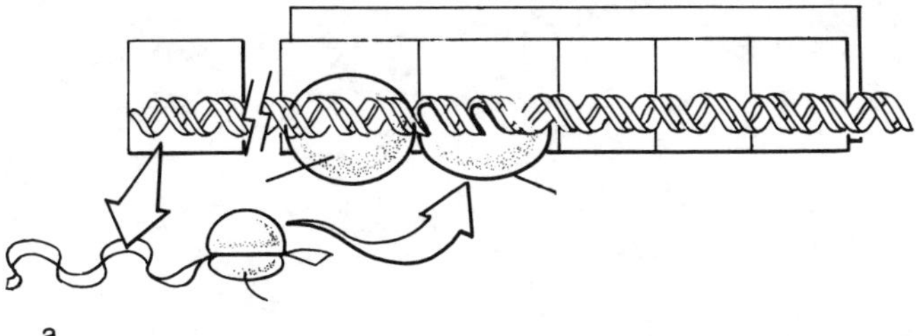

a

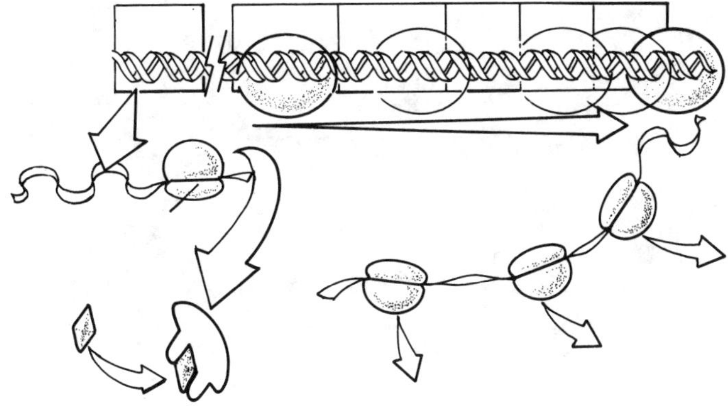

b

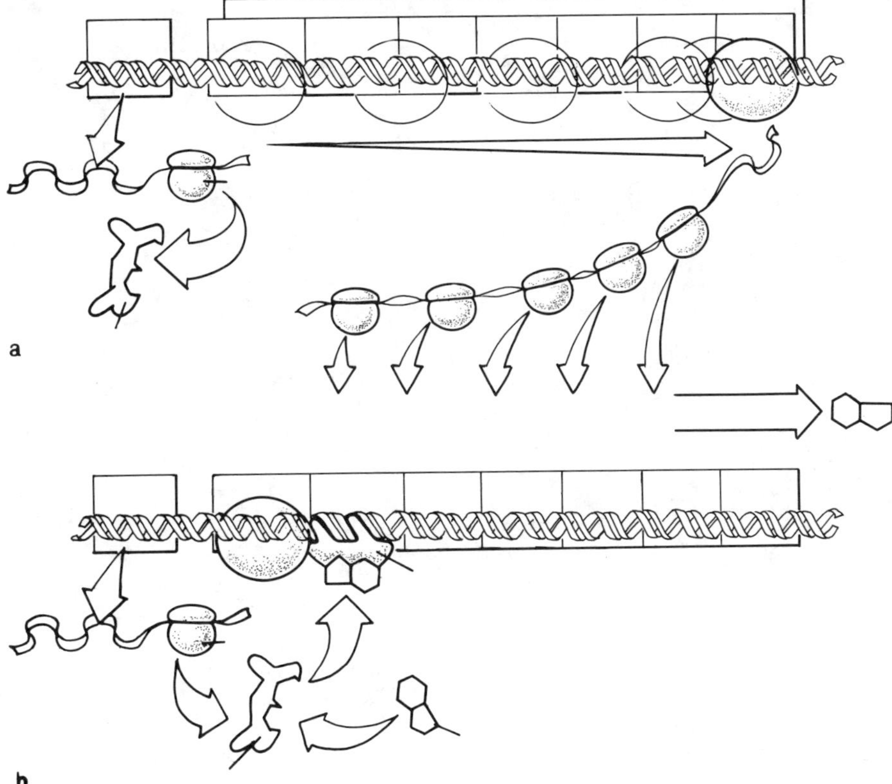

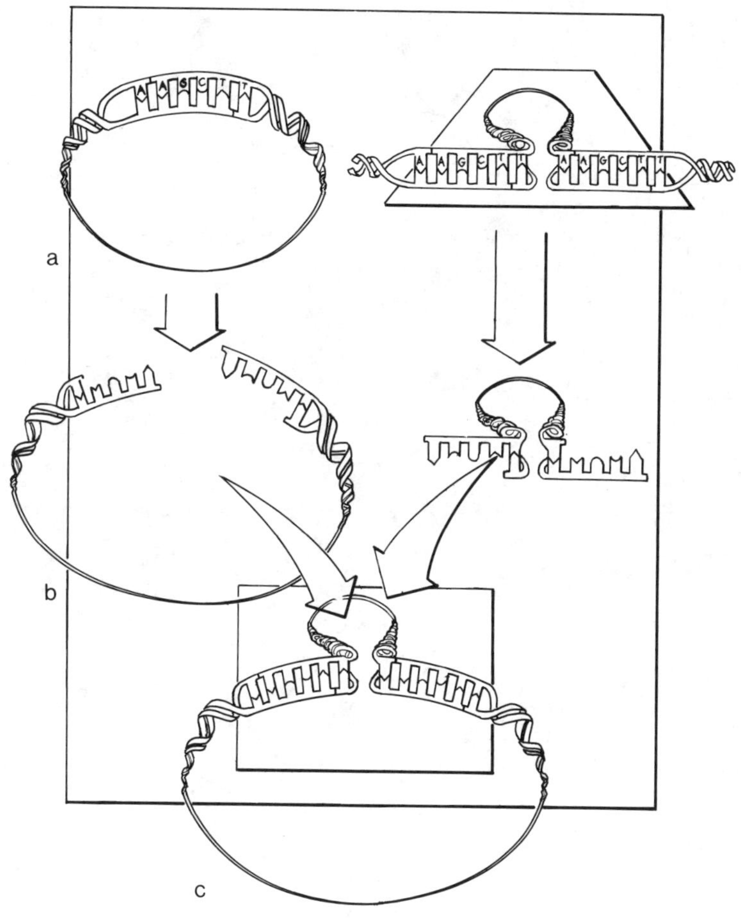
a
b
c

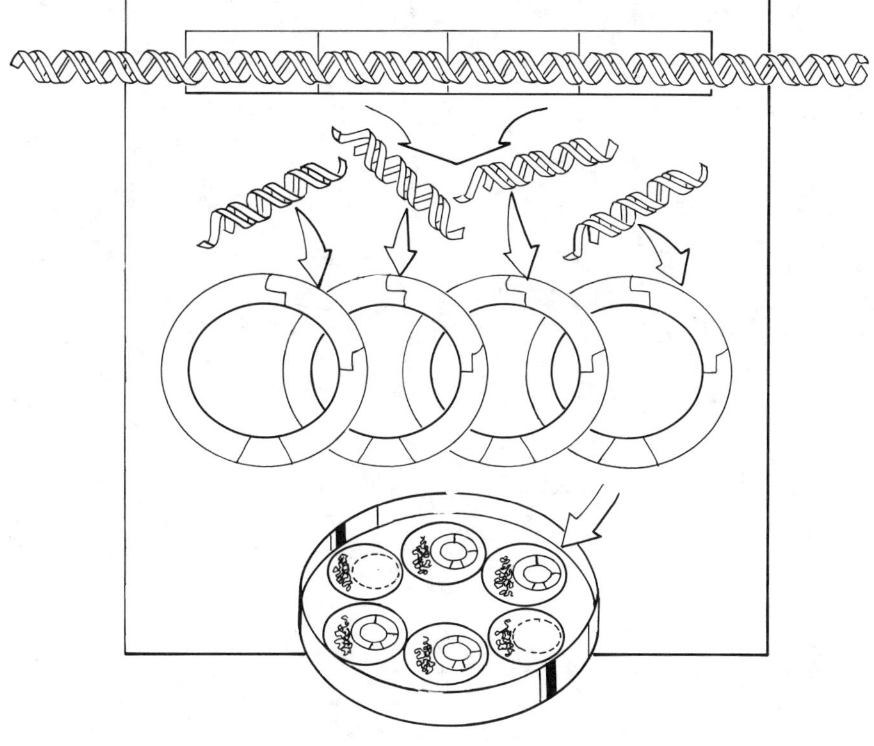

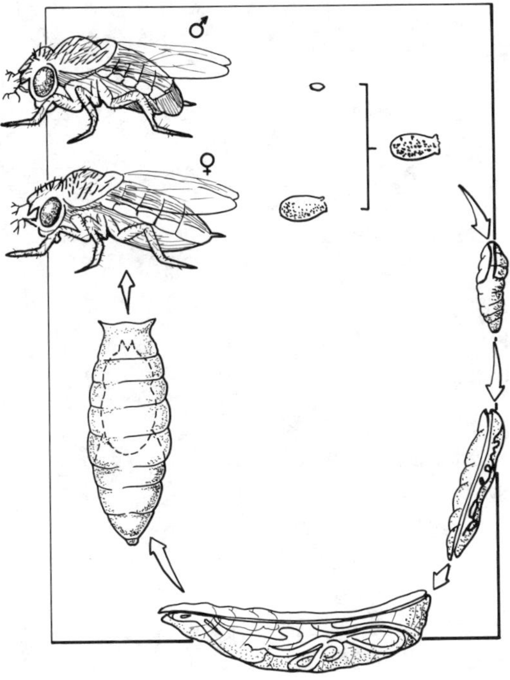

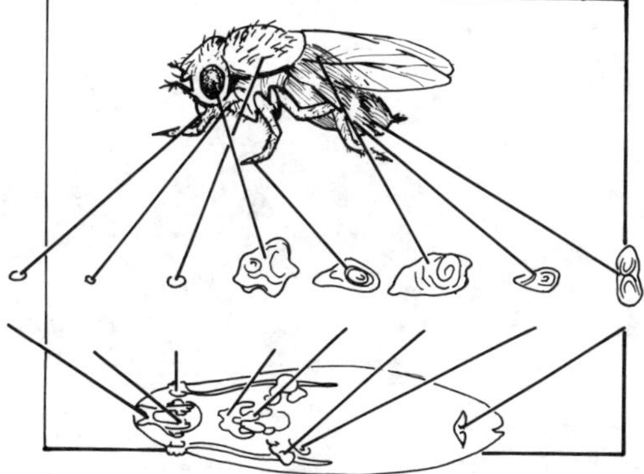

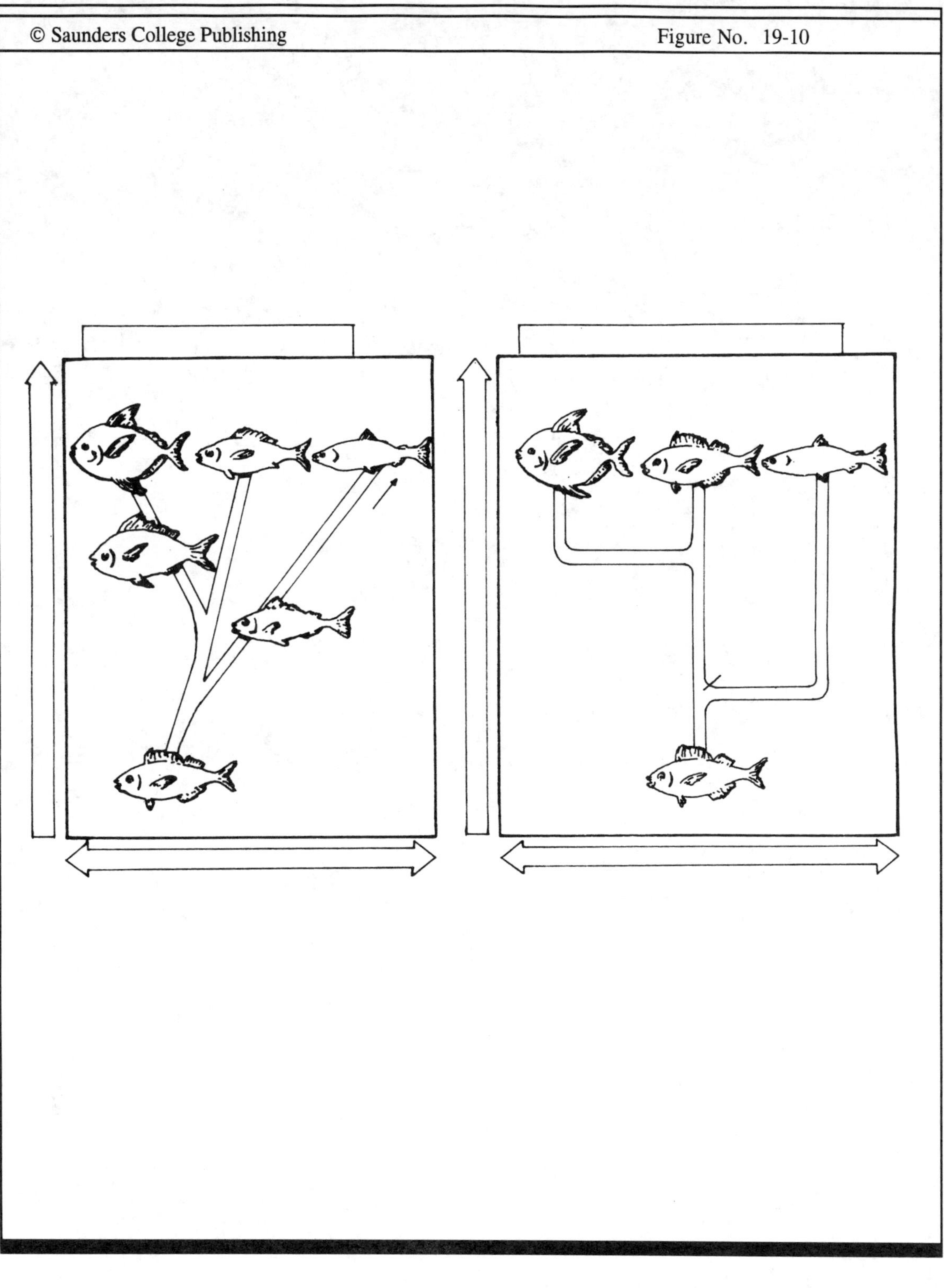

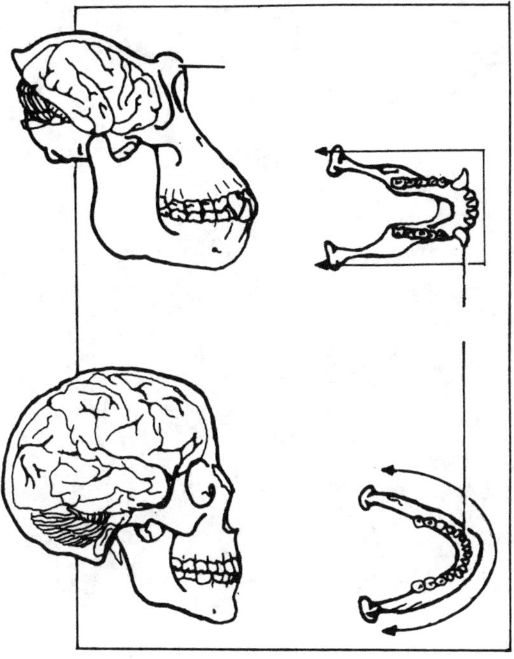

Figure No. 24-6b

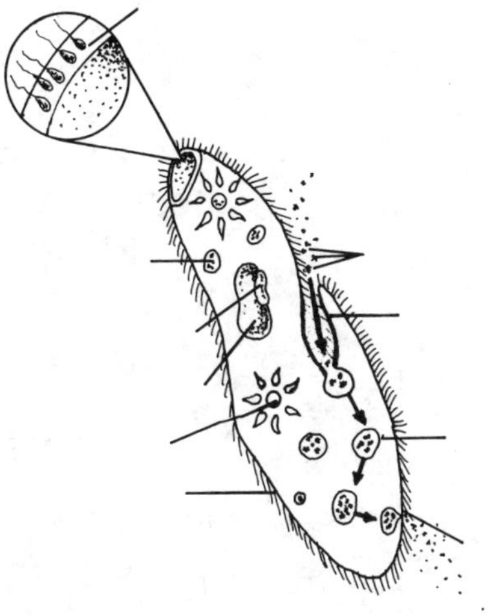

 Figure No. 24-10b

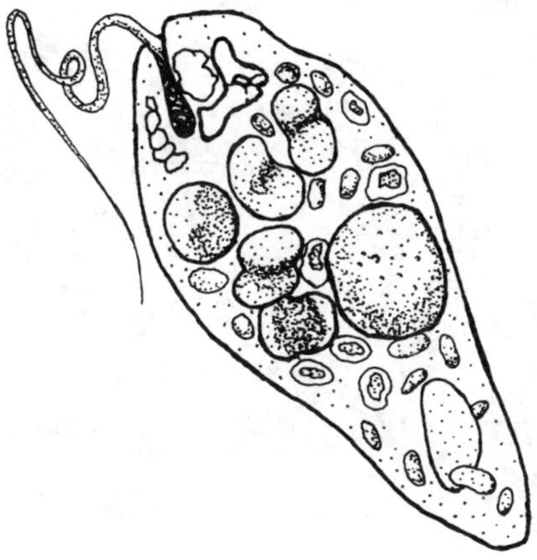

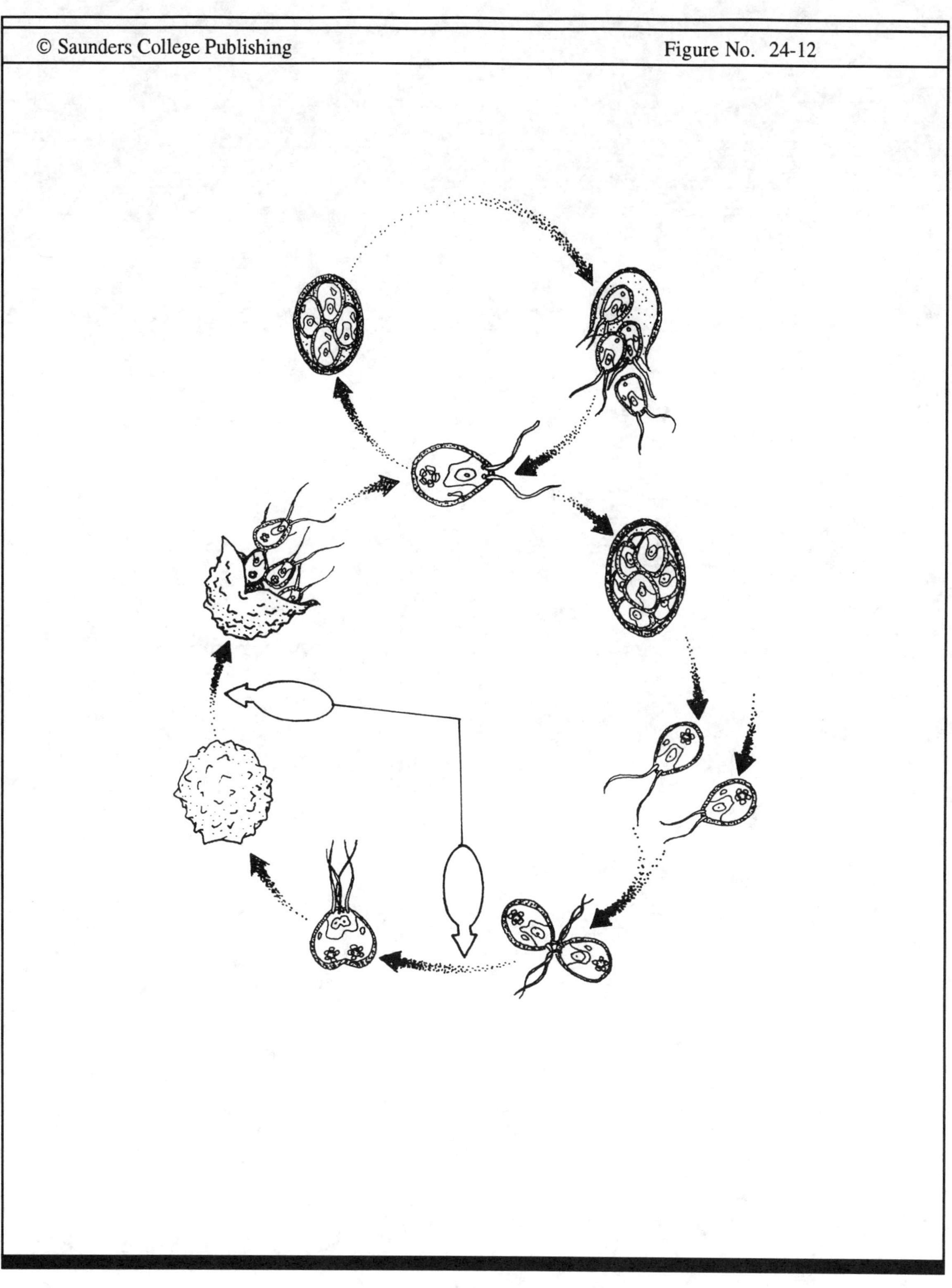

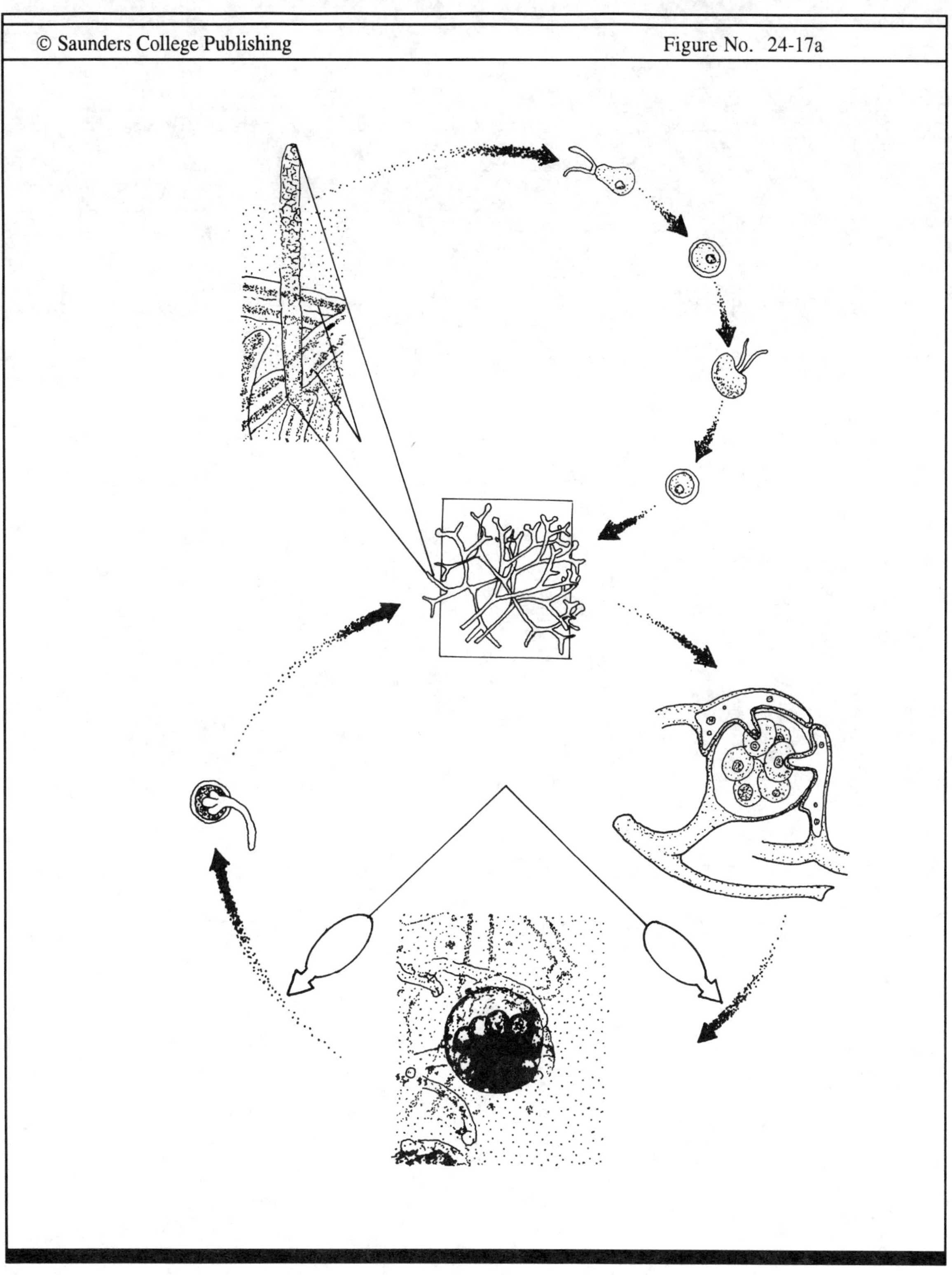

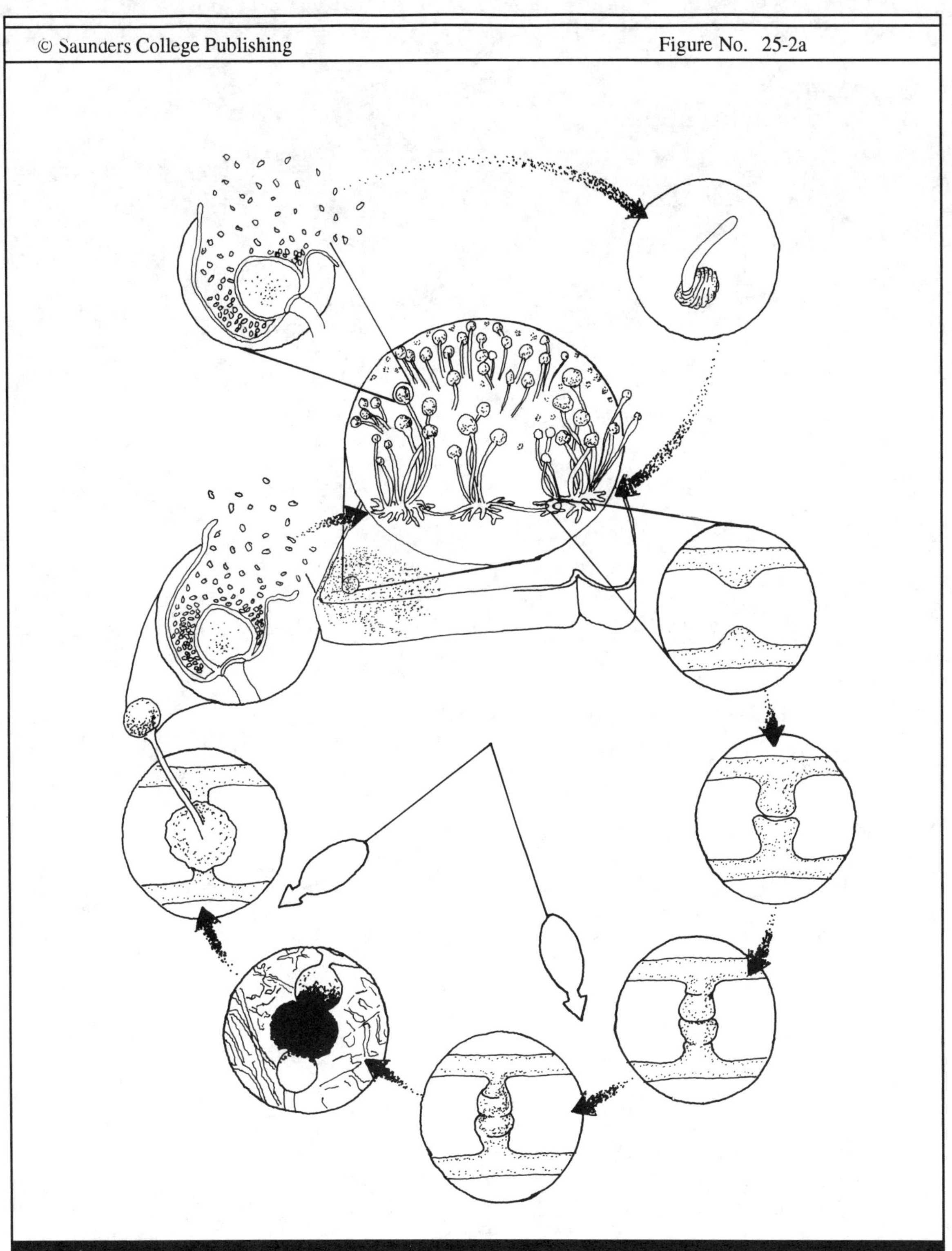

 Figure No. 26-5

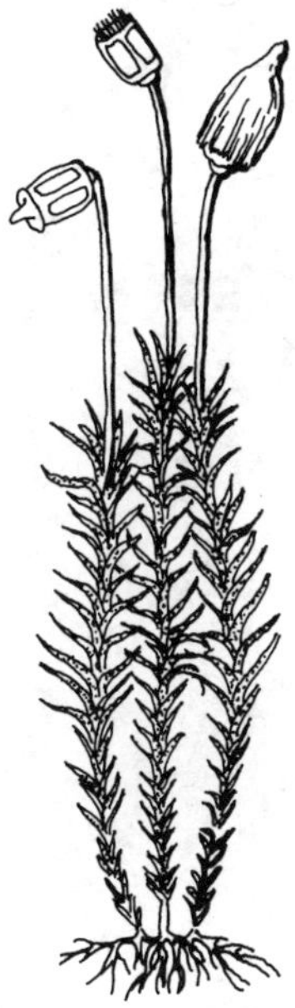

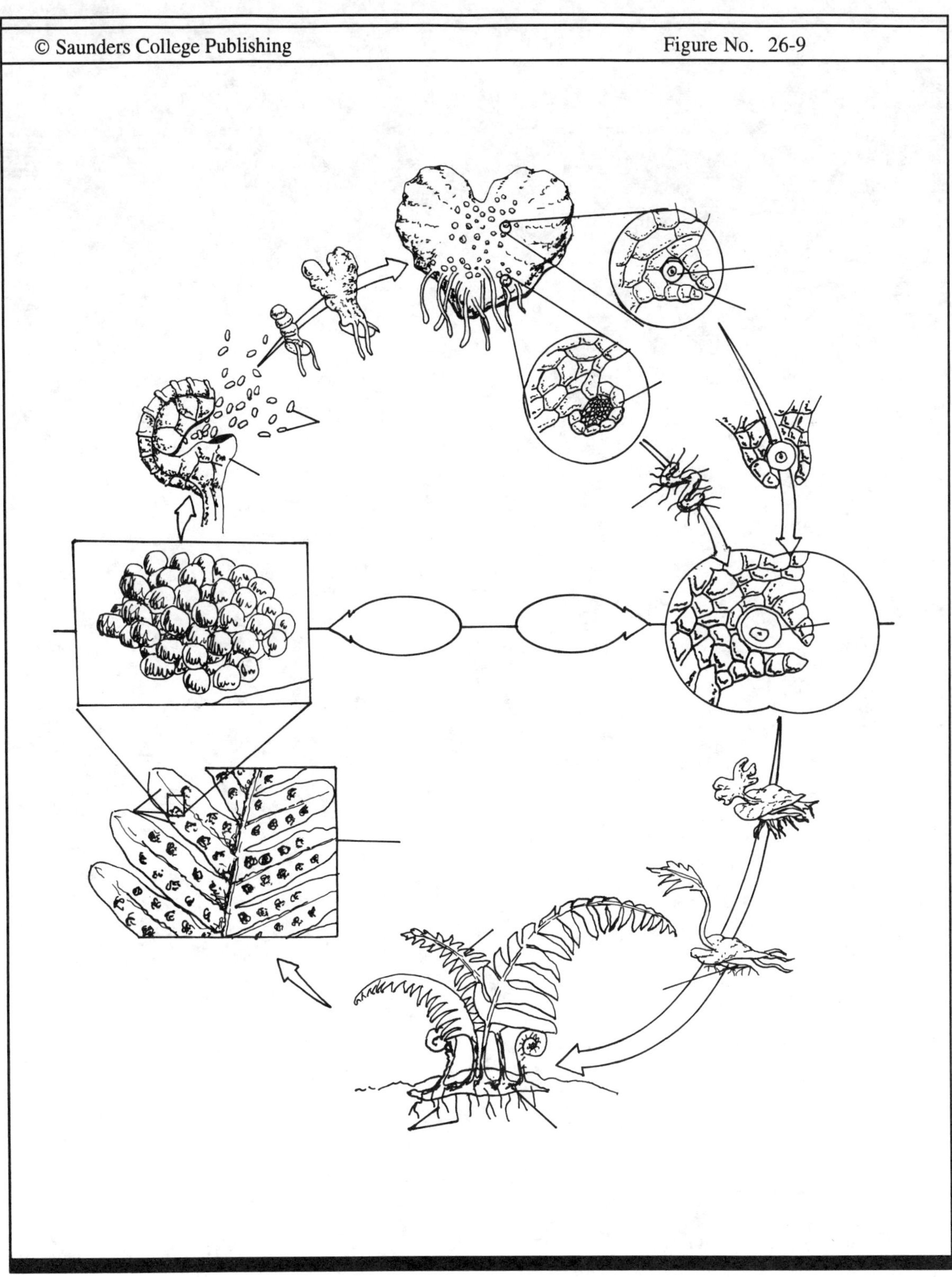

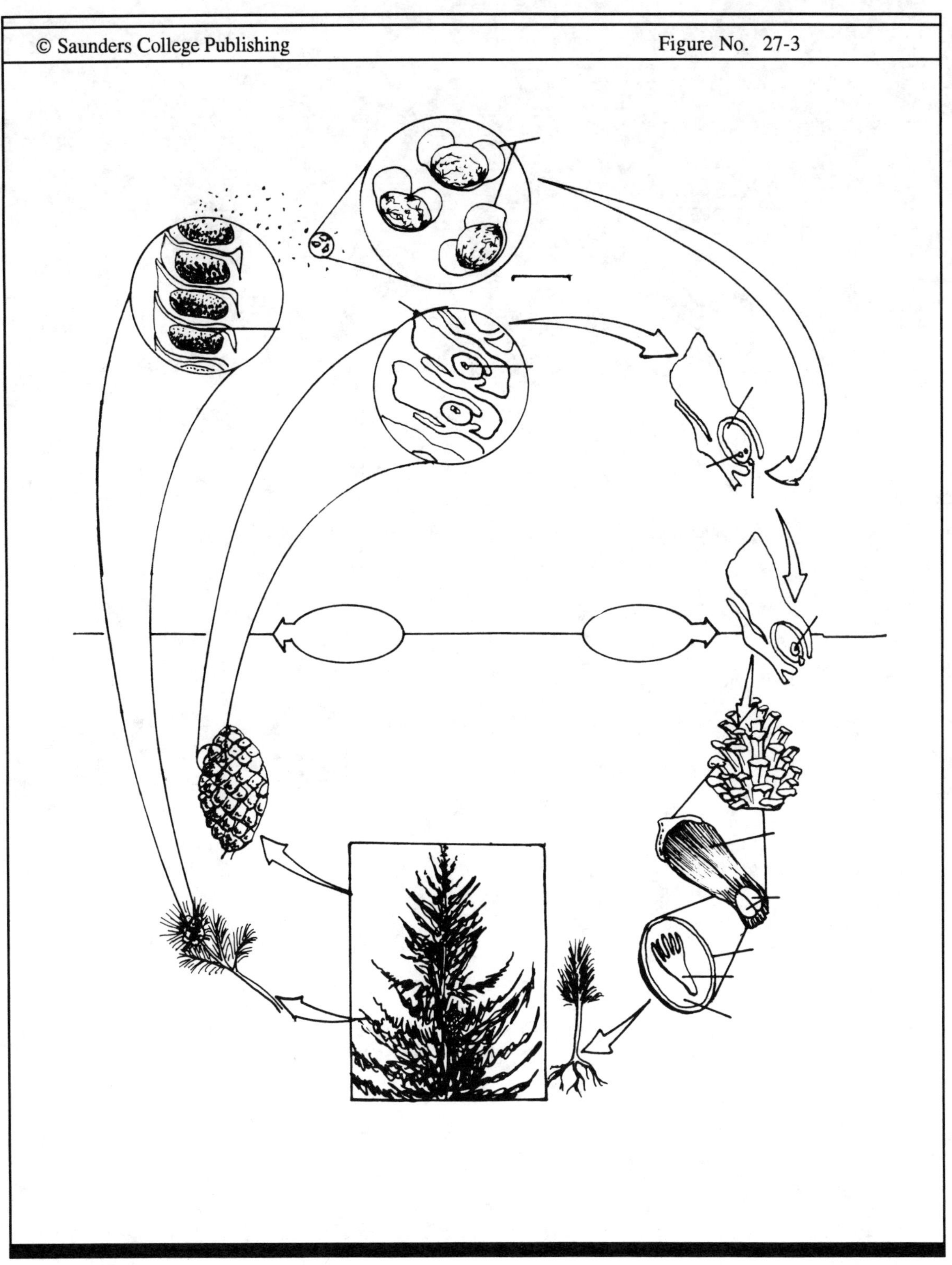

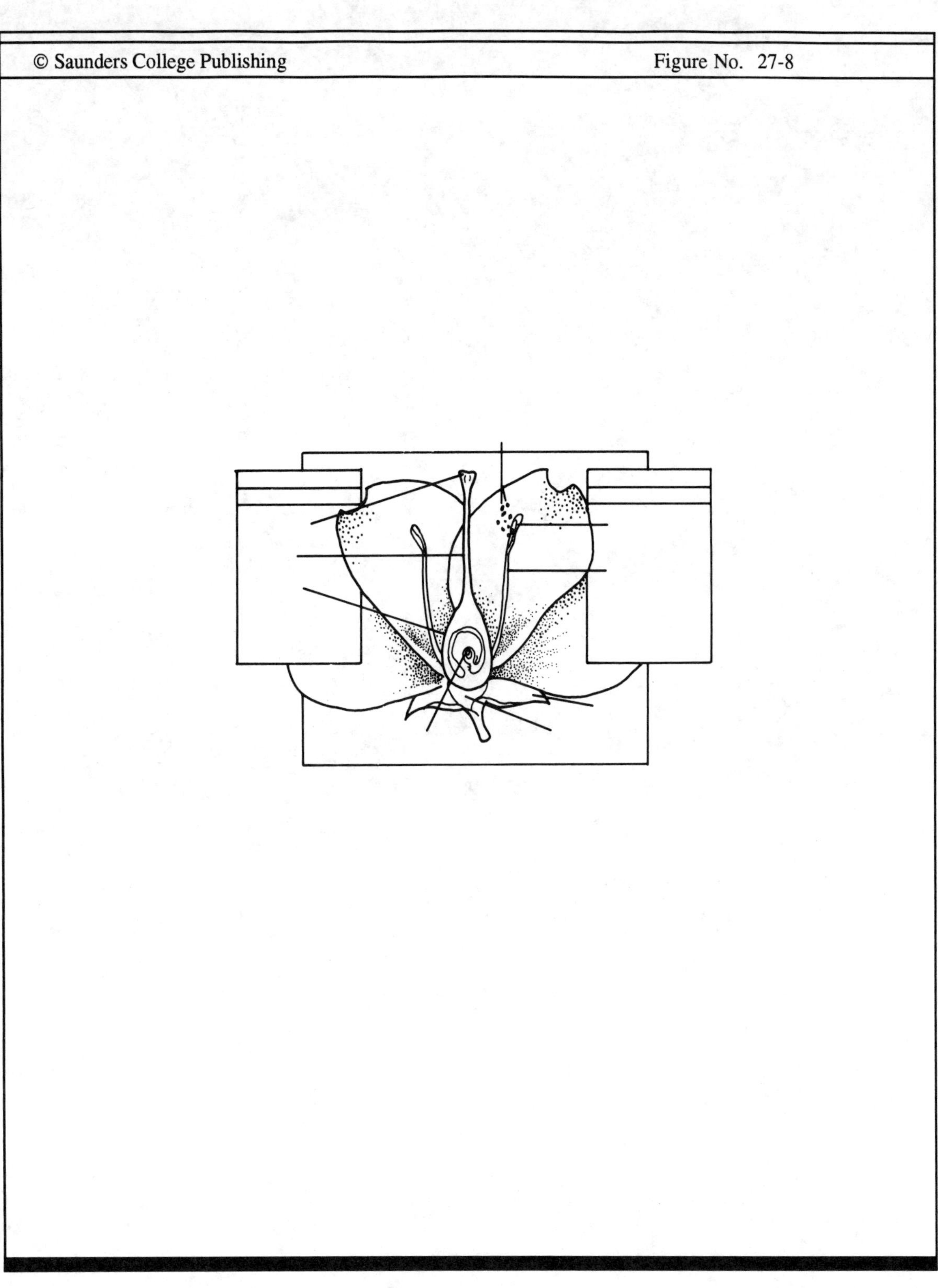

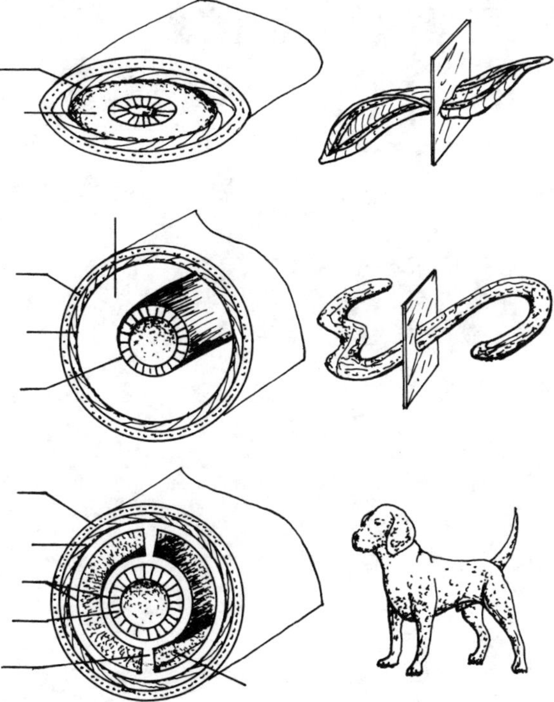

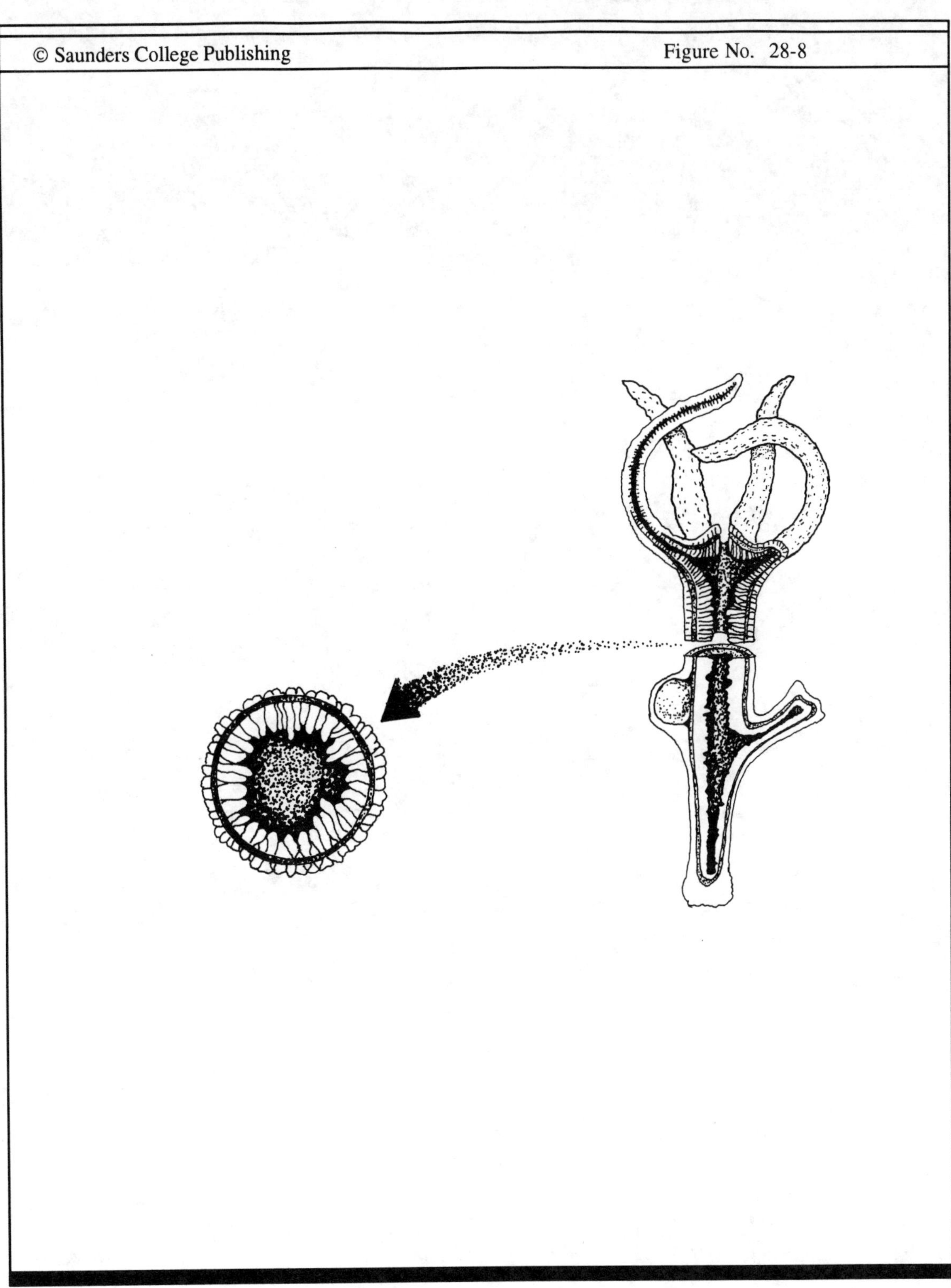

 Figure No. 28-12

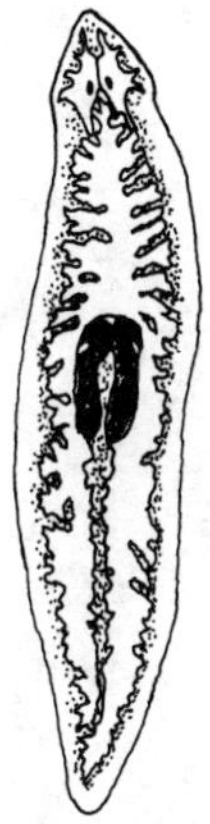

 Figure No. 28-19

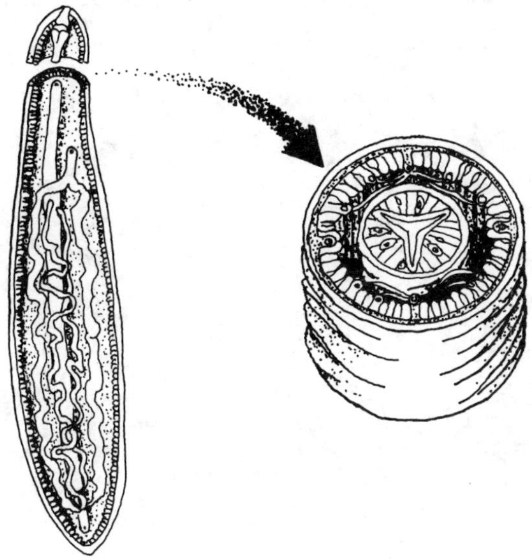

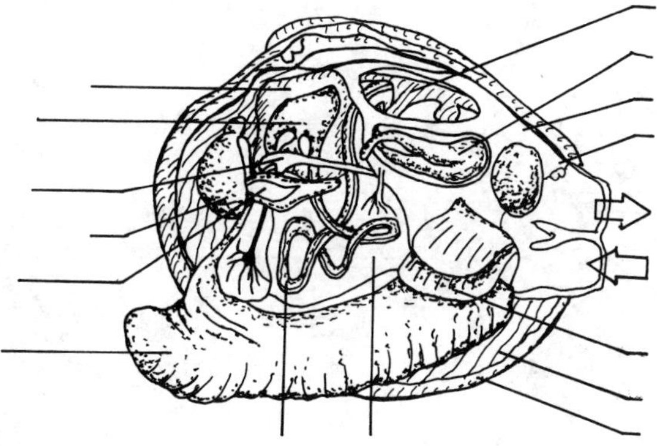

 Figure No. 30-14

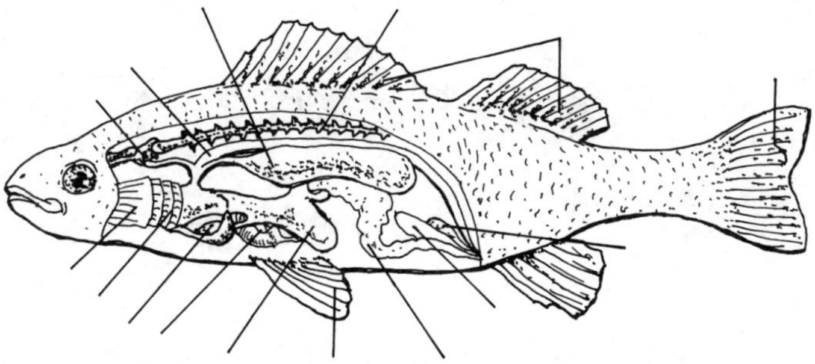

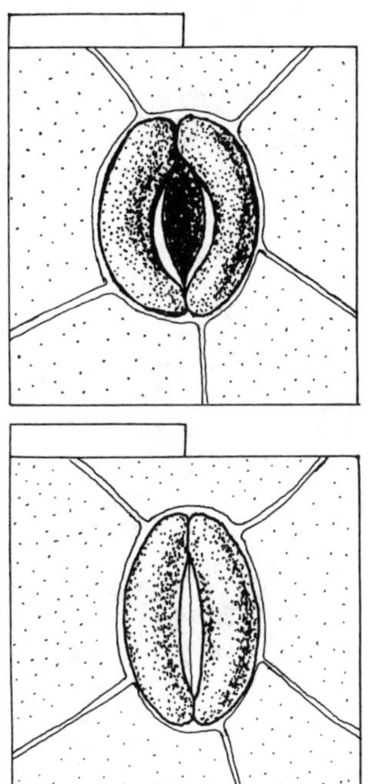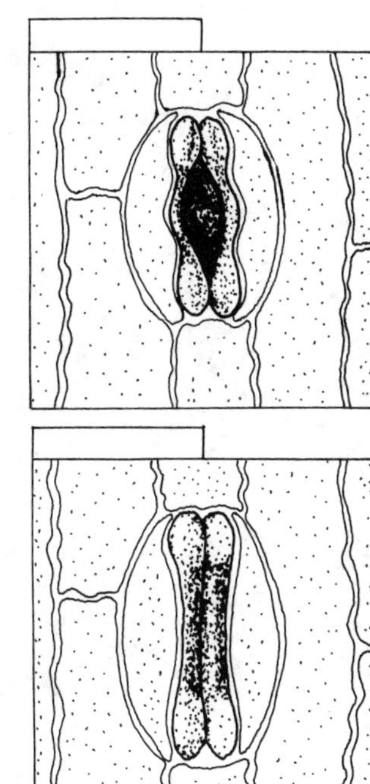

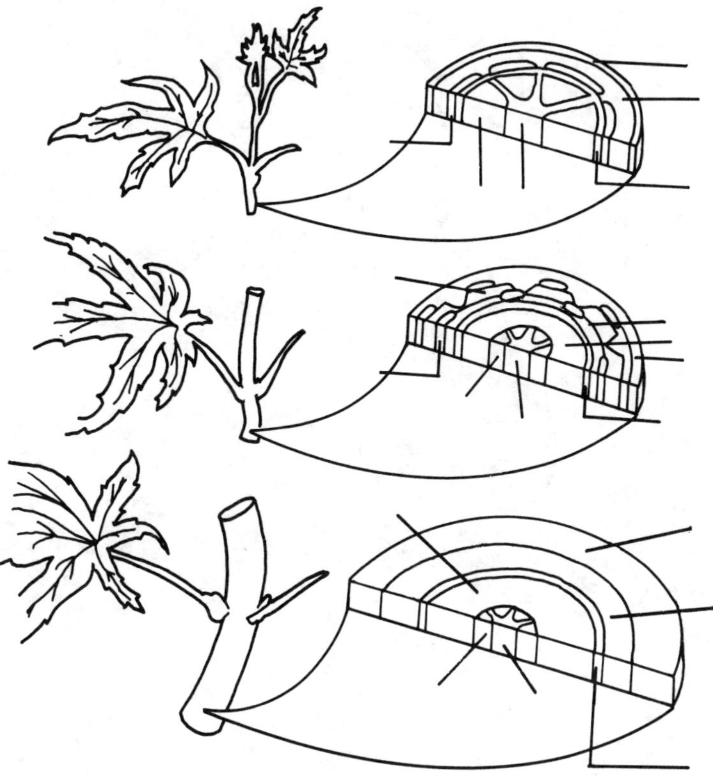

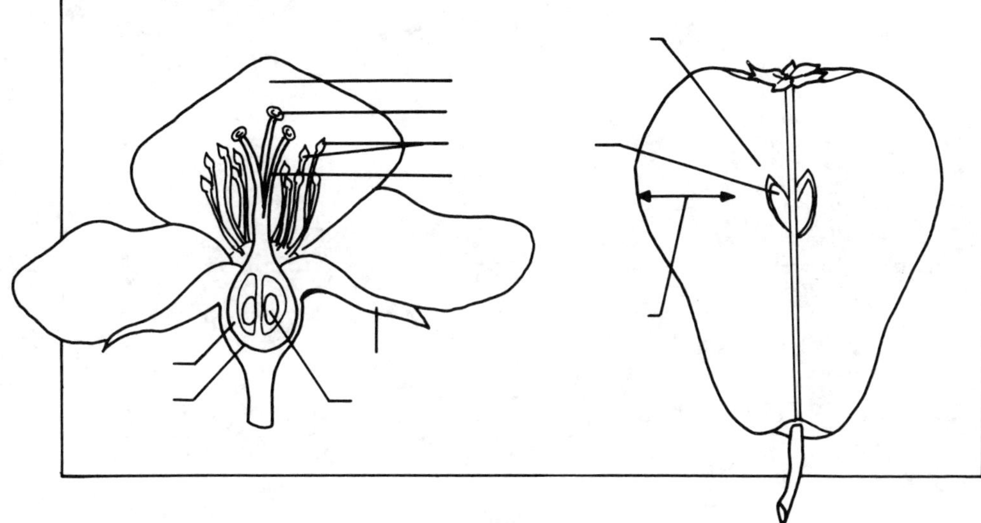

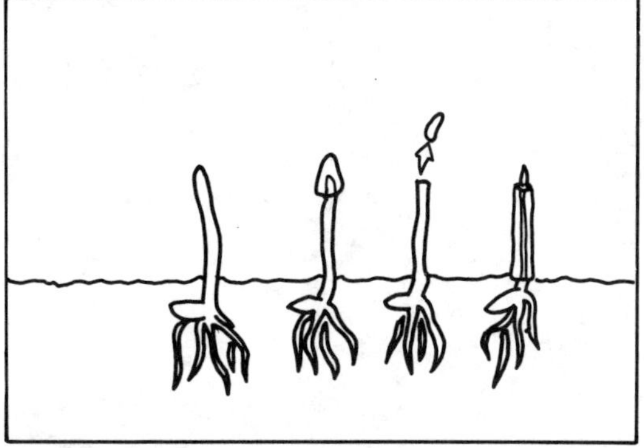

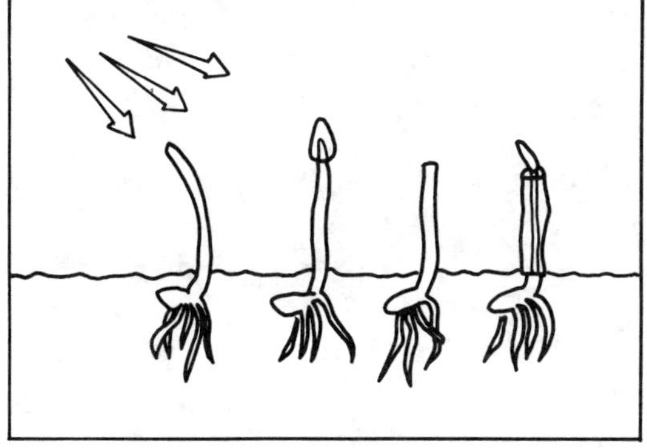

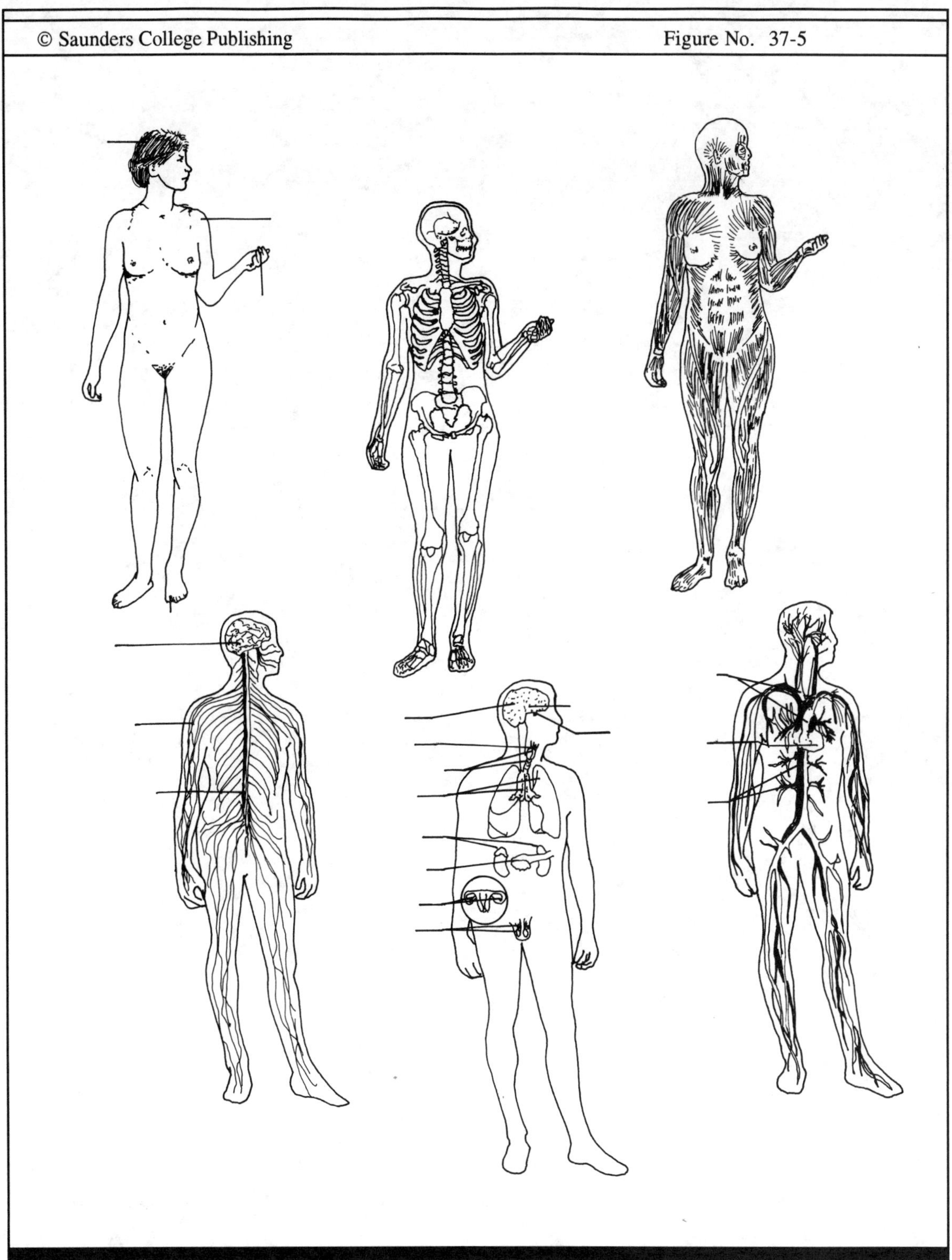

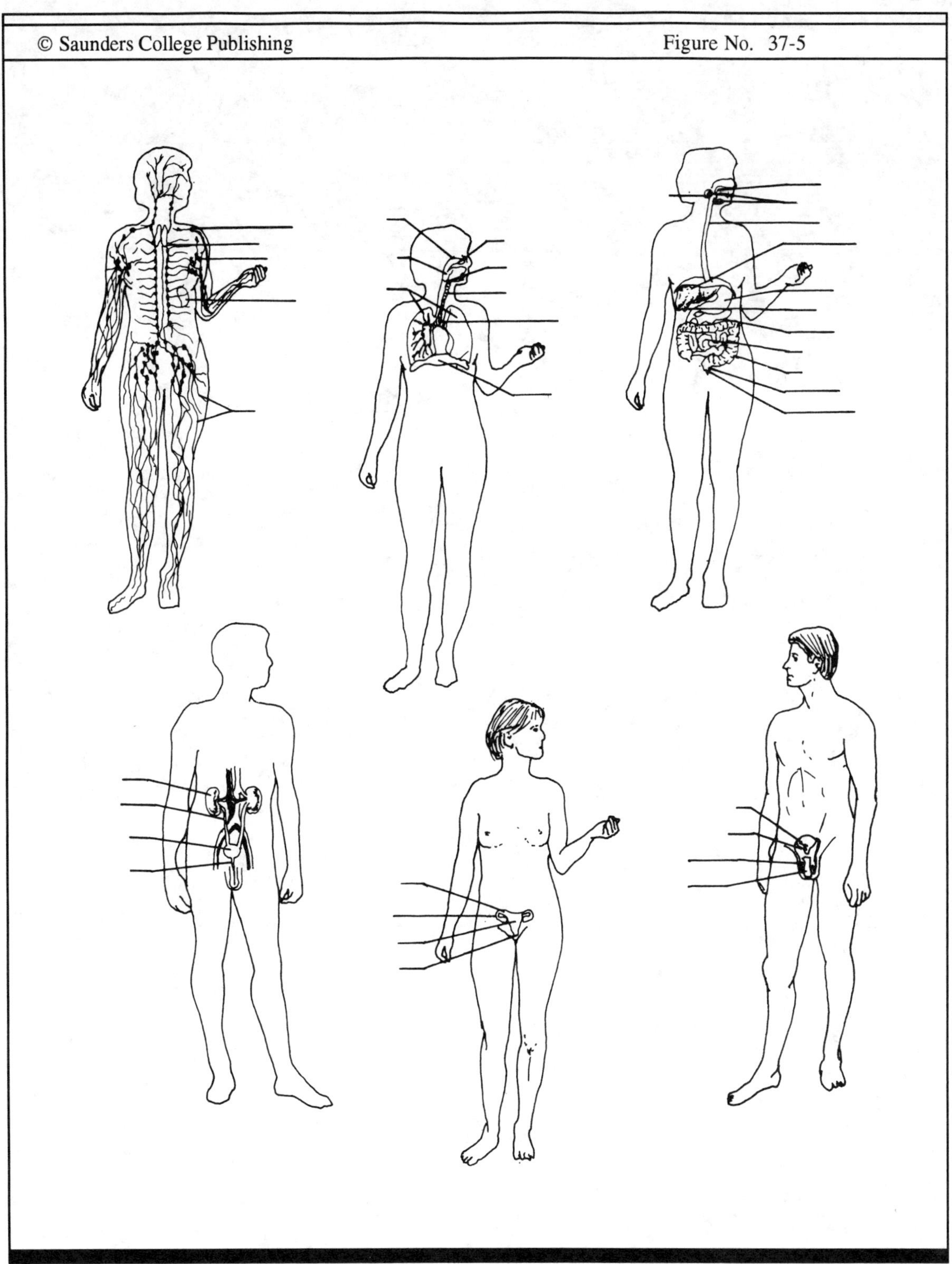

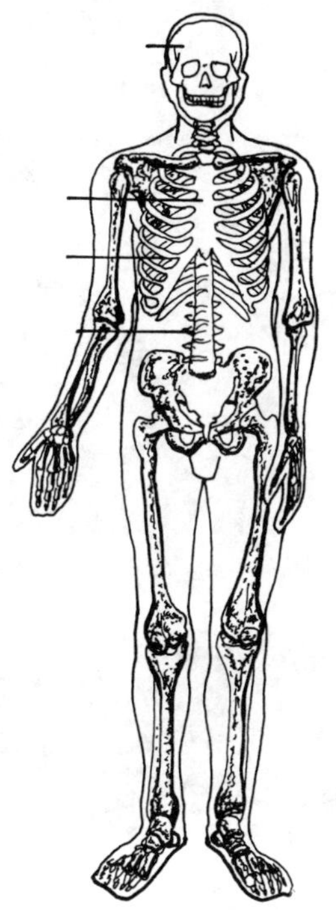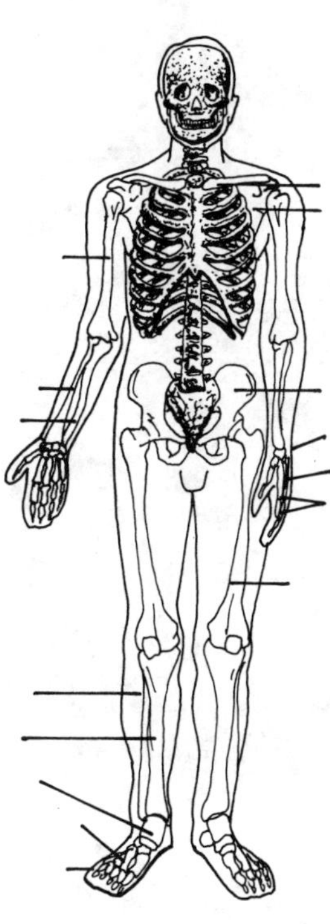

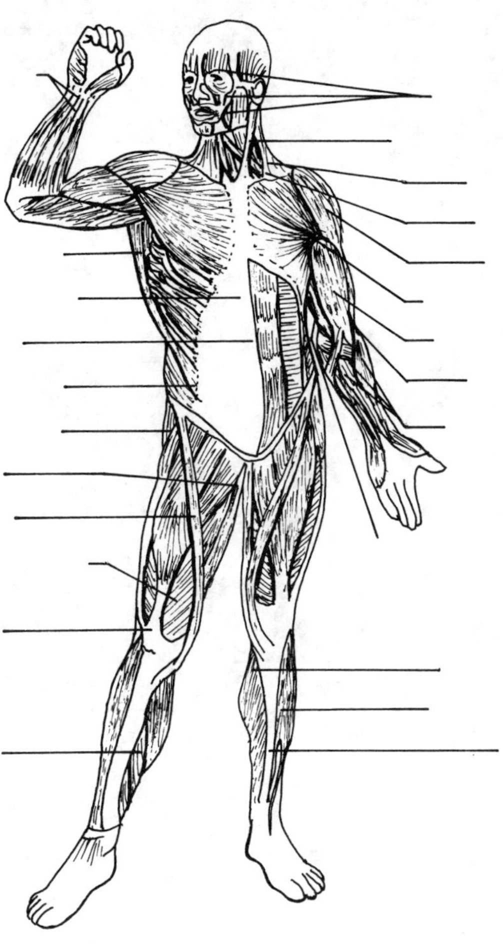

 Figure No. 38-15

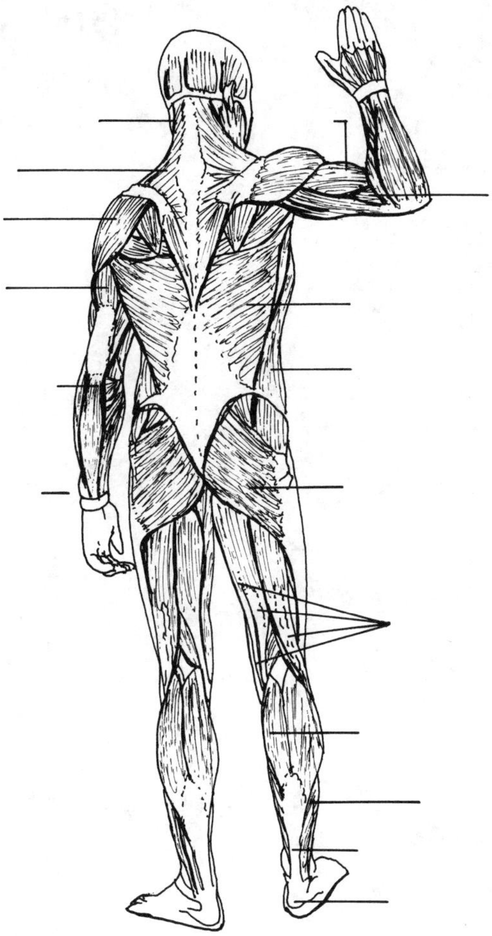

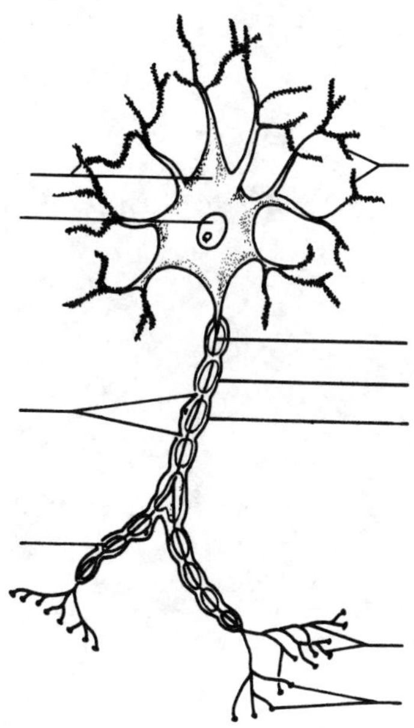

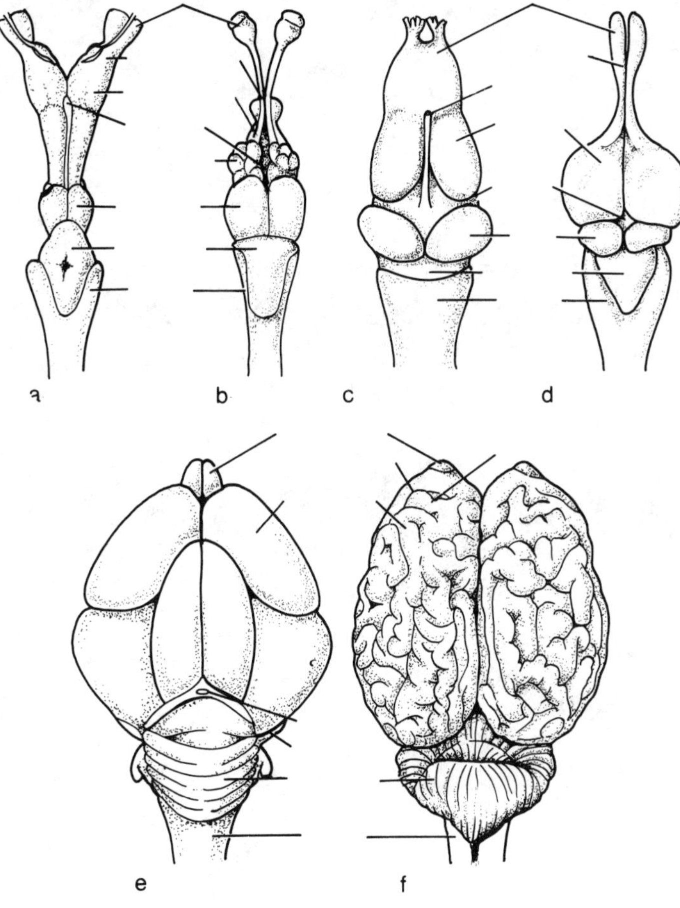
a
b
c
d
e
f

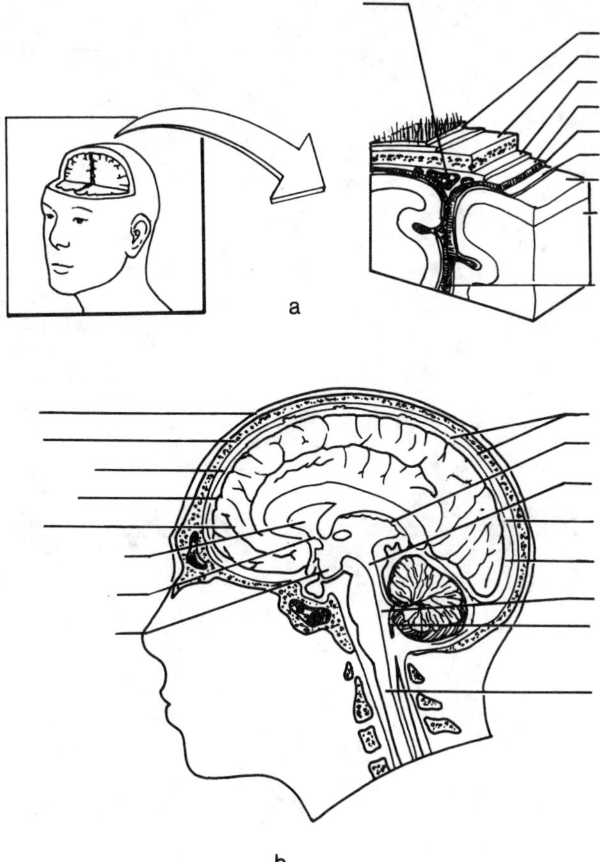

a
b

 Figure No. 40-8

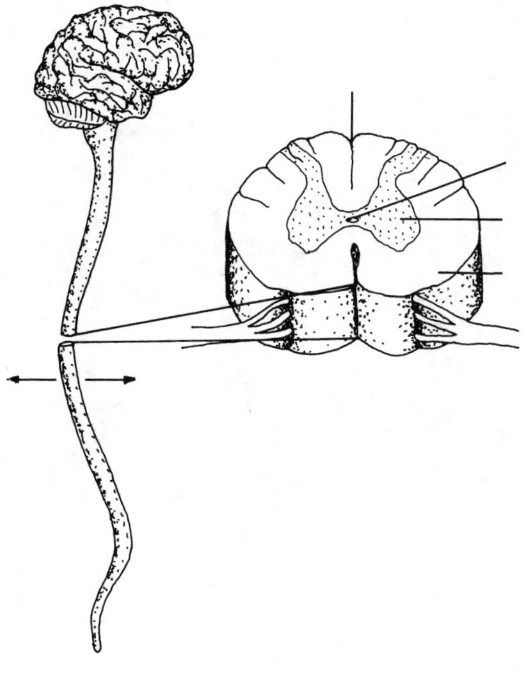

Figure No. 40-9

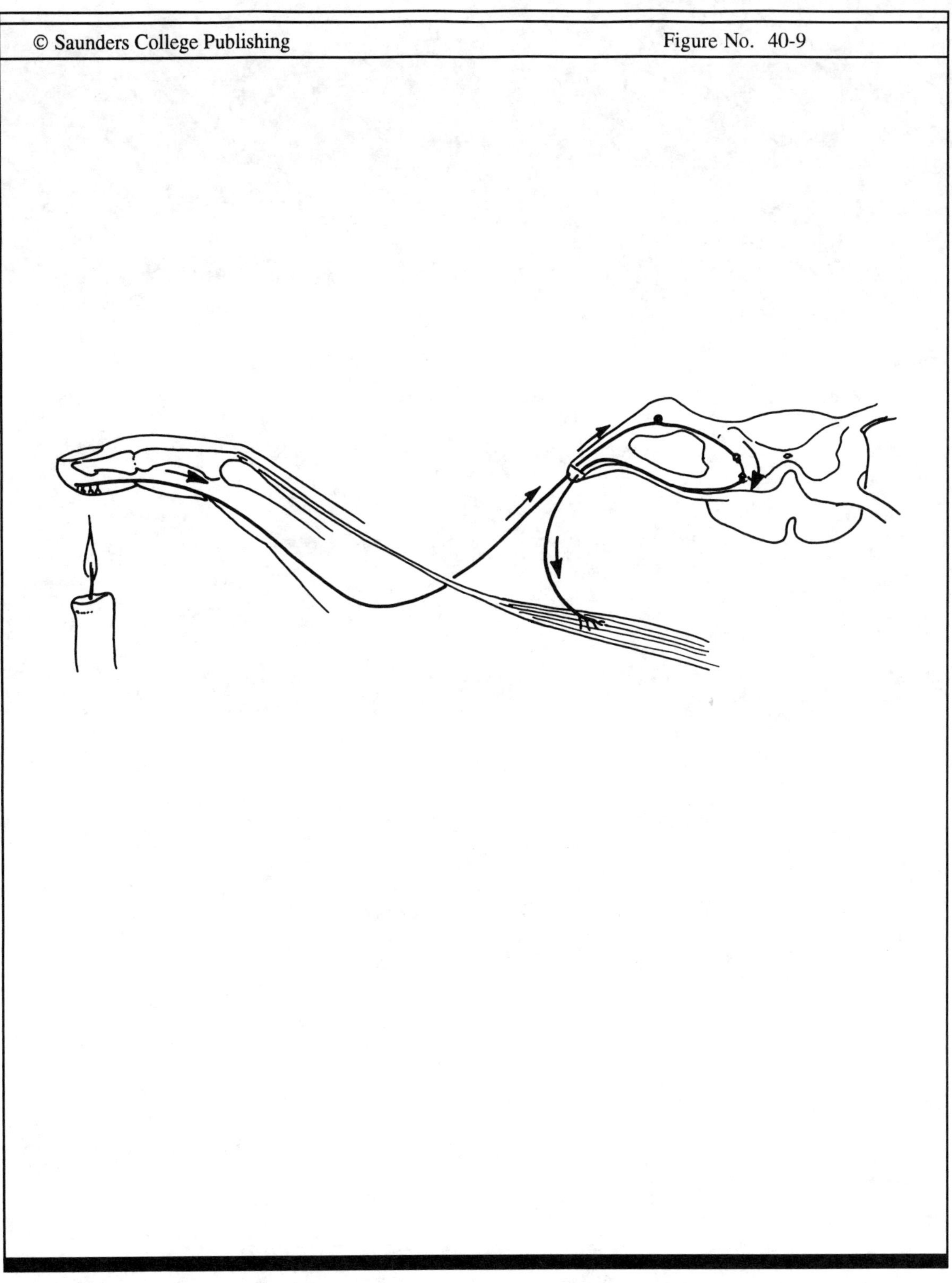

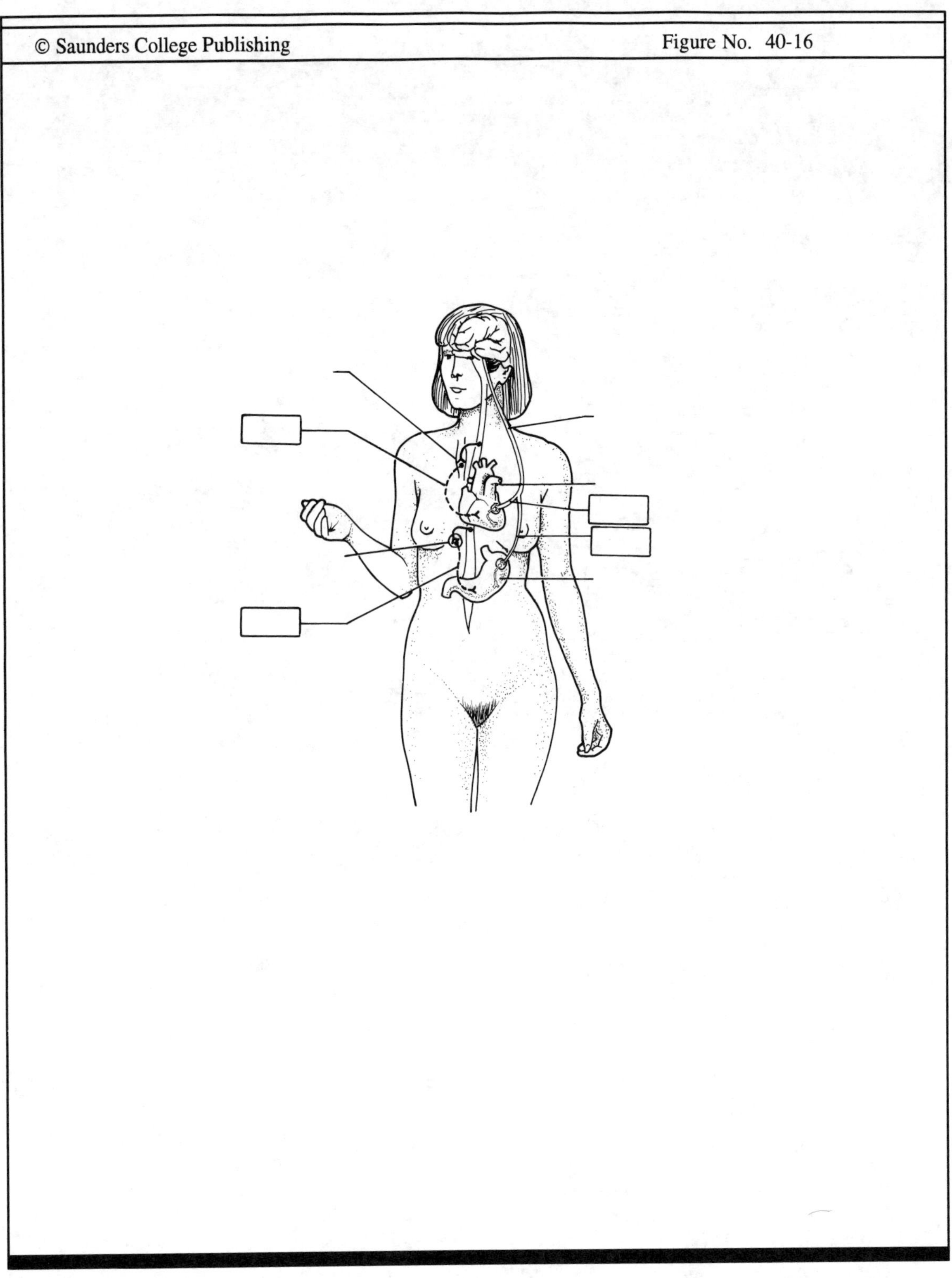

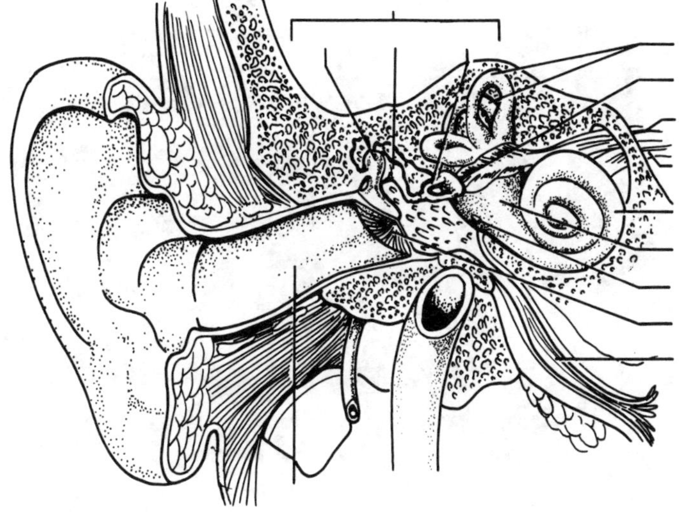

 Figure No. 41-16

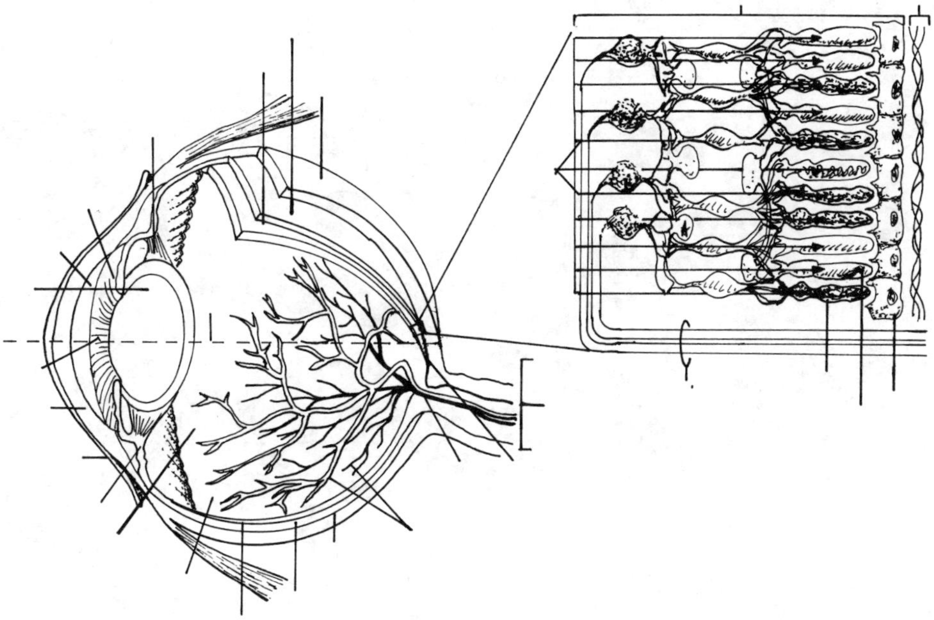

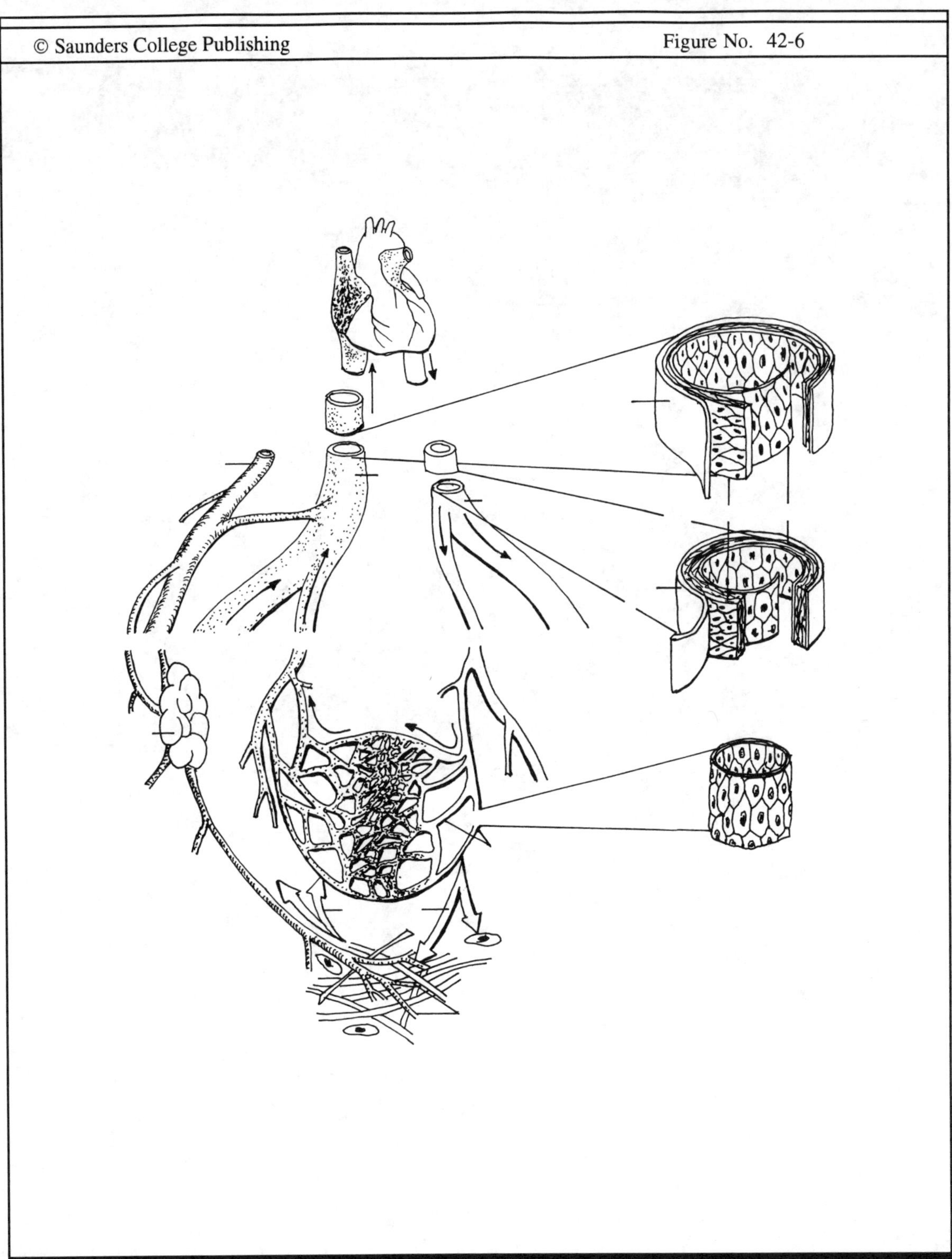

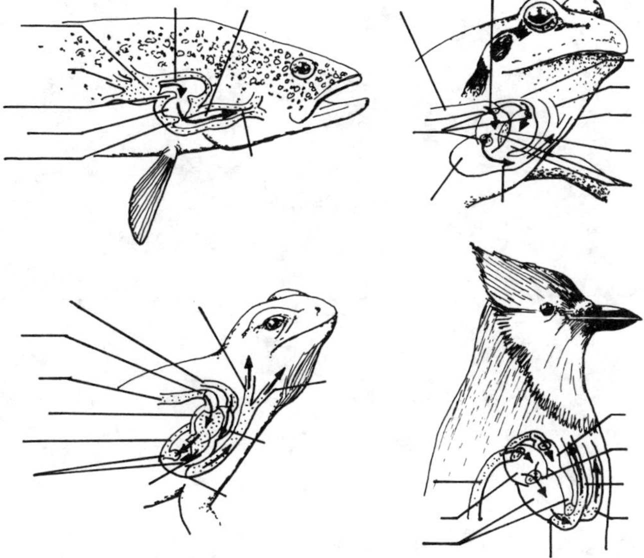

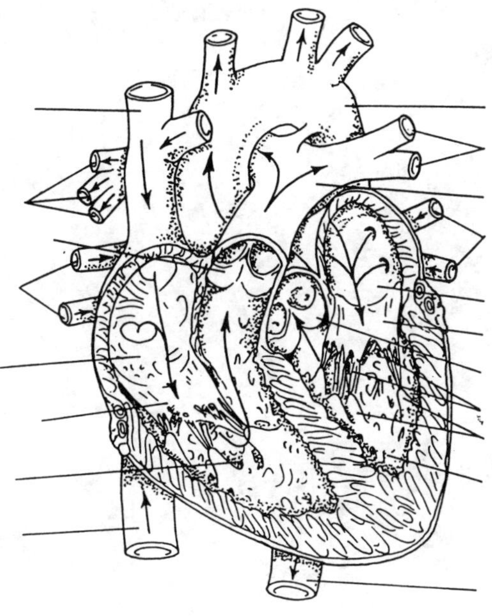

Figure No. 42-16

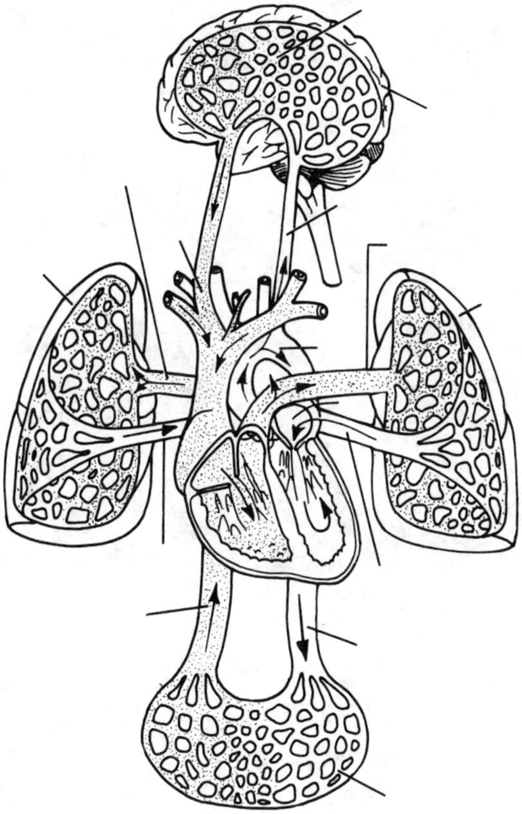

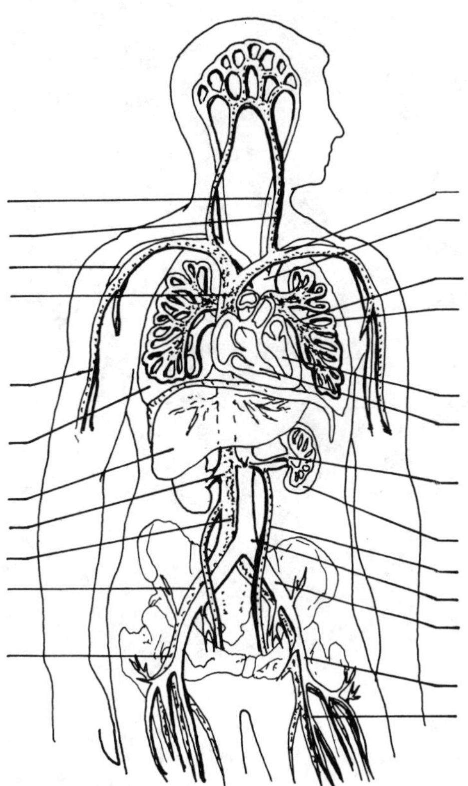

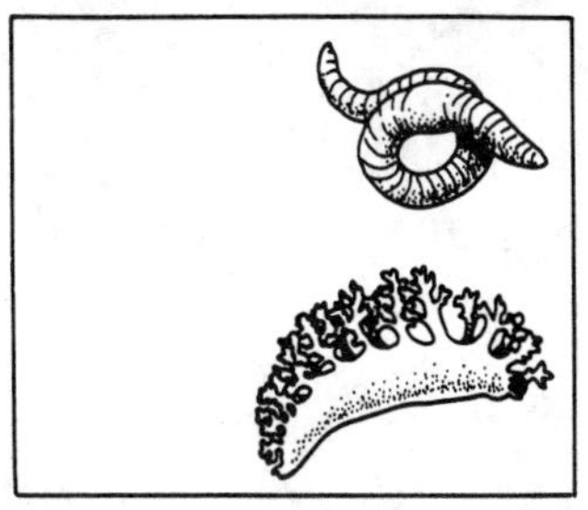

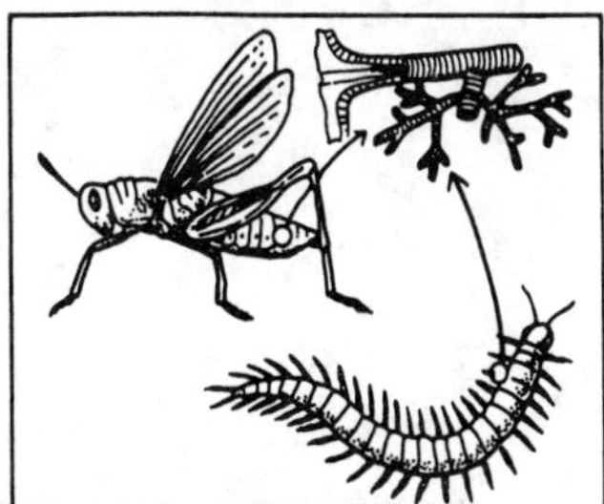

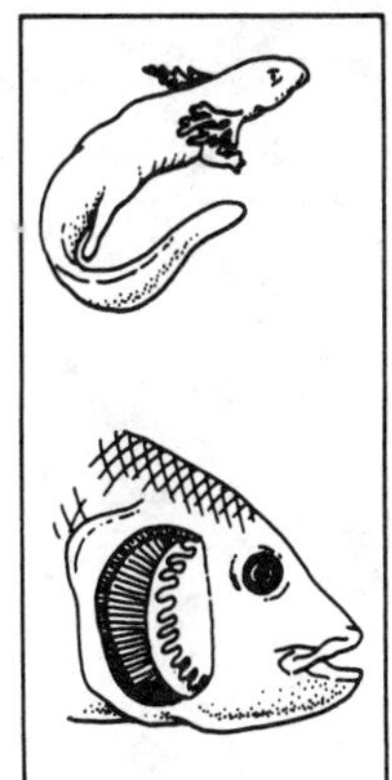

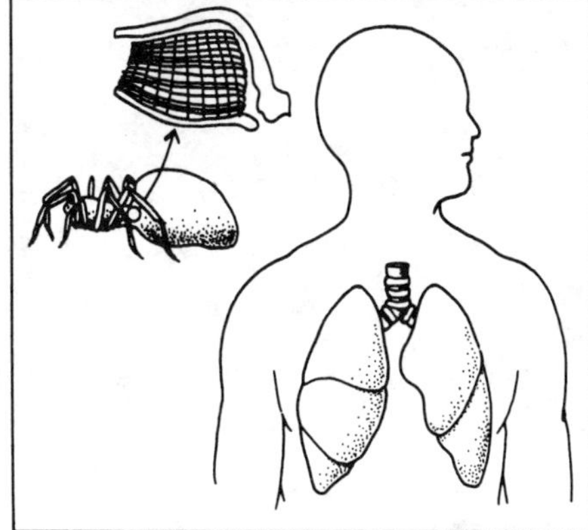

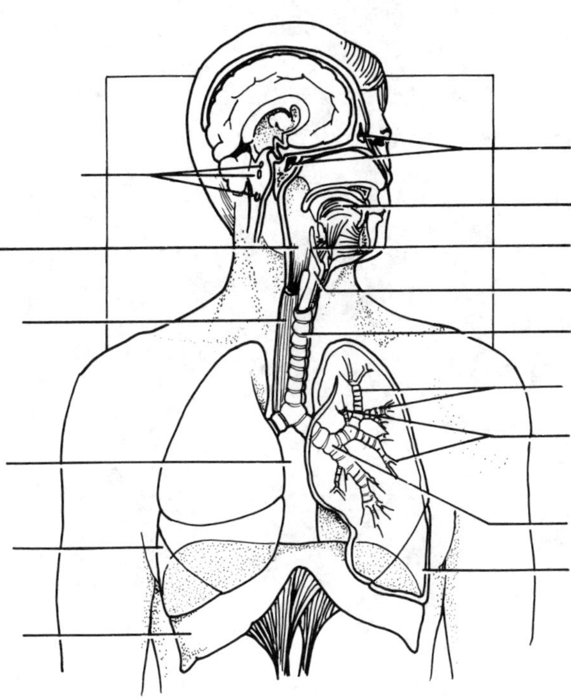

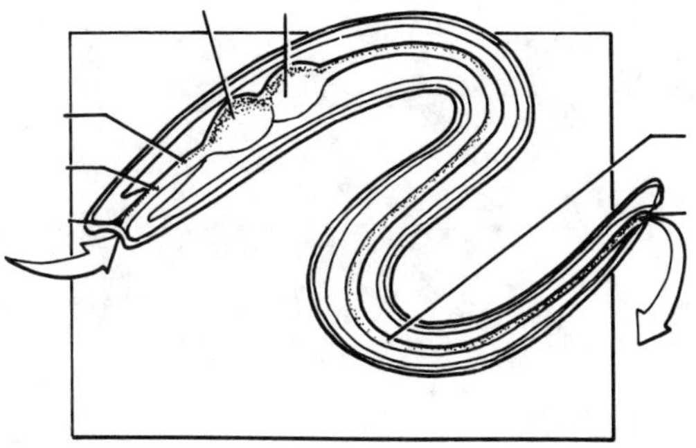

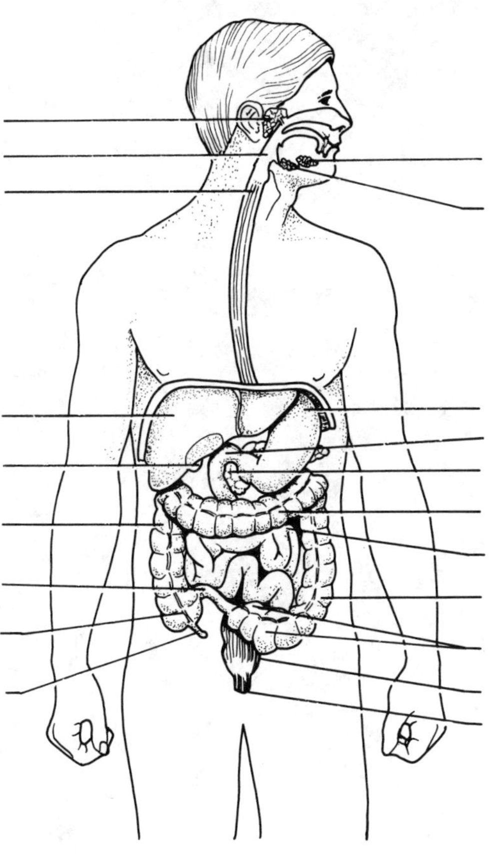

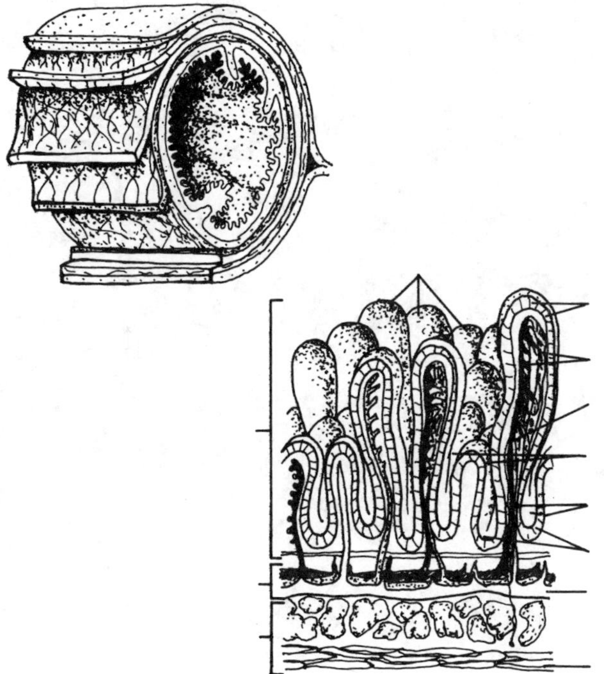

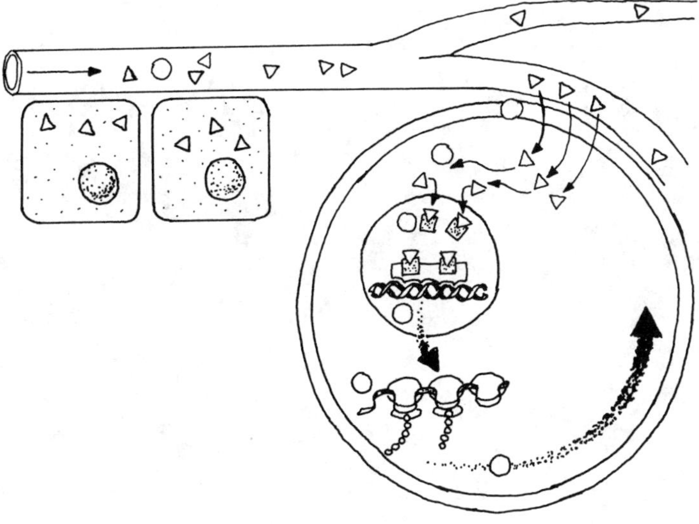

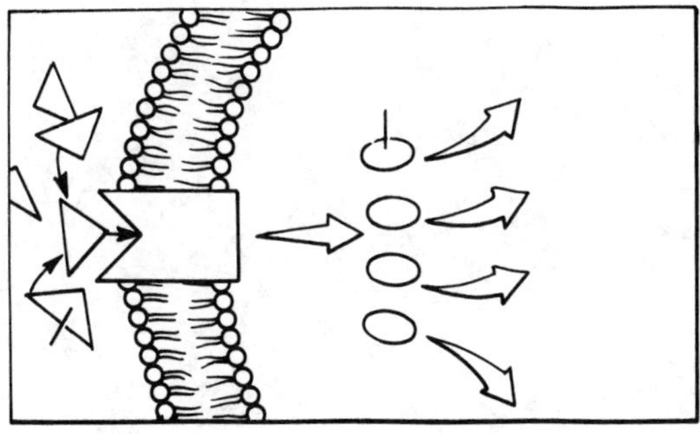

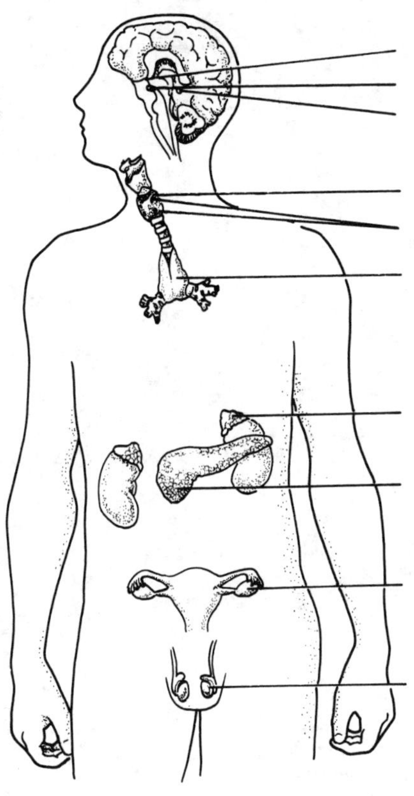

Figure No. 47-12

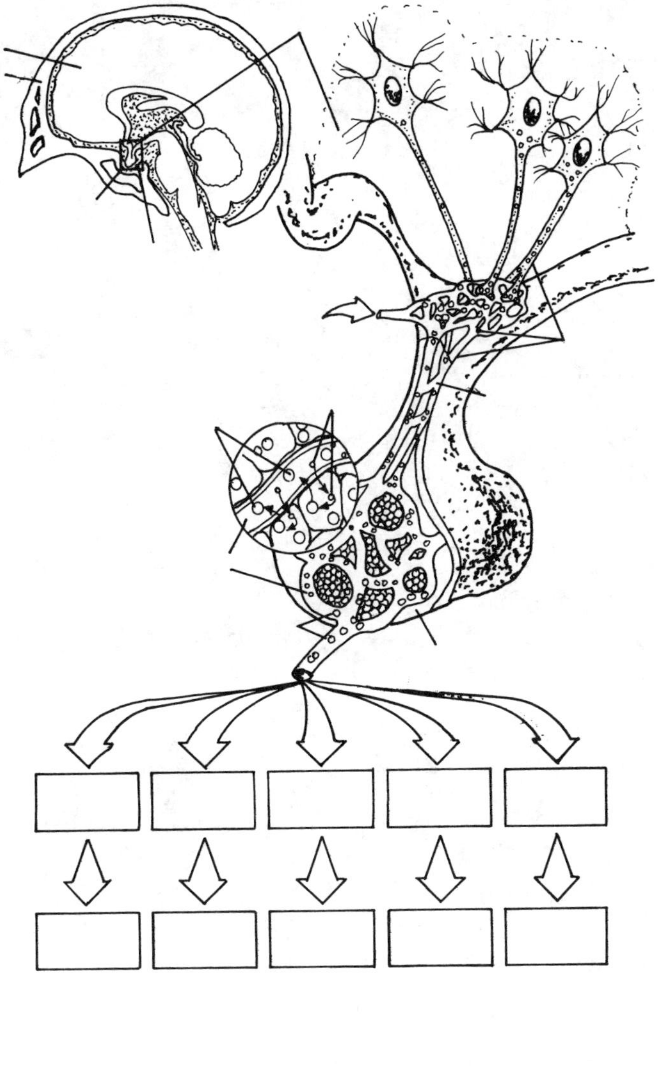

Figure No. 47-15

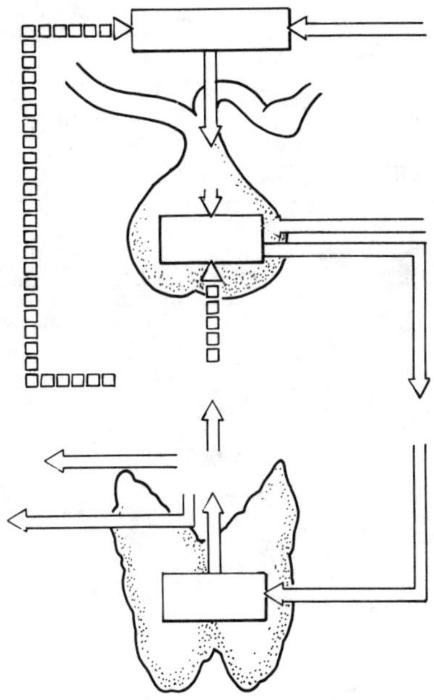

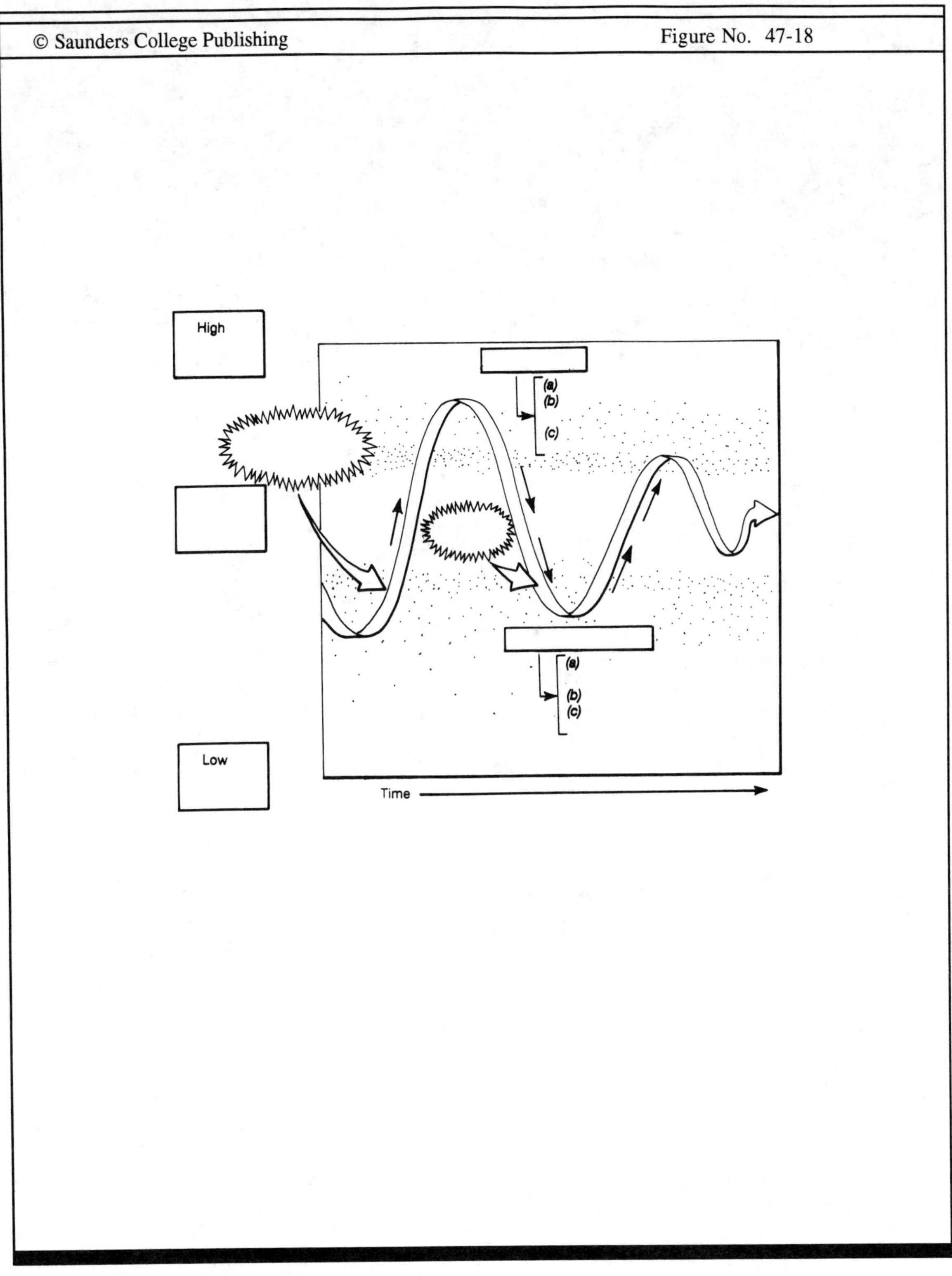

High
Low
Time
(a)
(b)
(c)
(a)
(b)
(c)

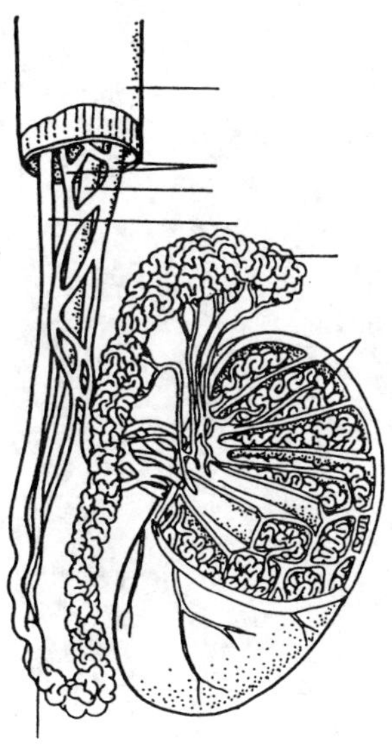

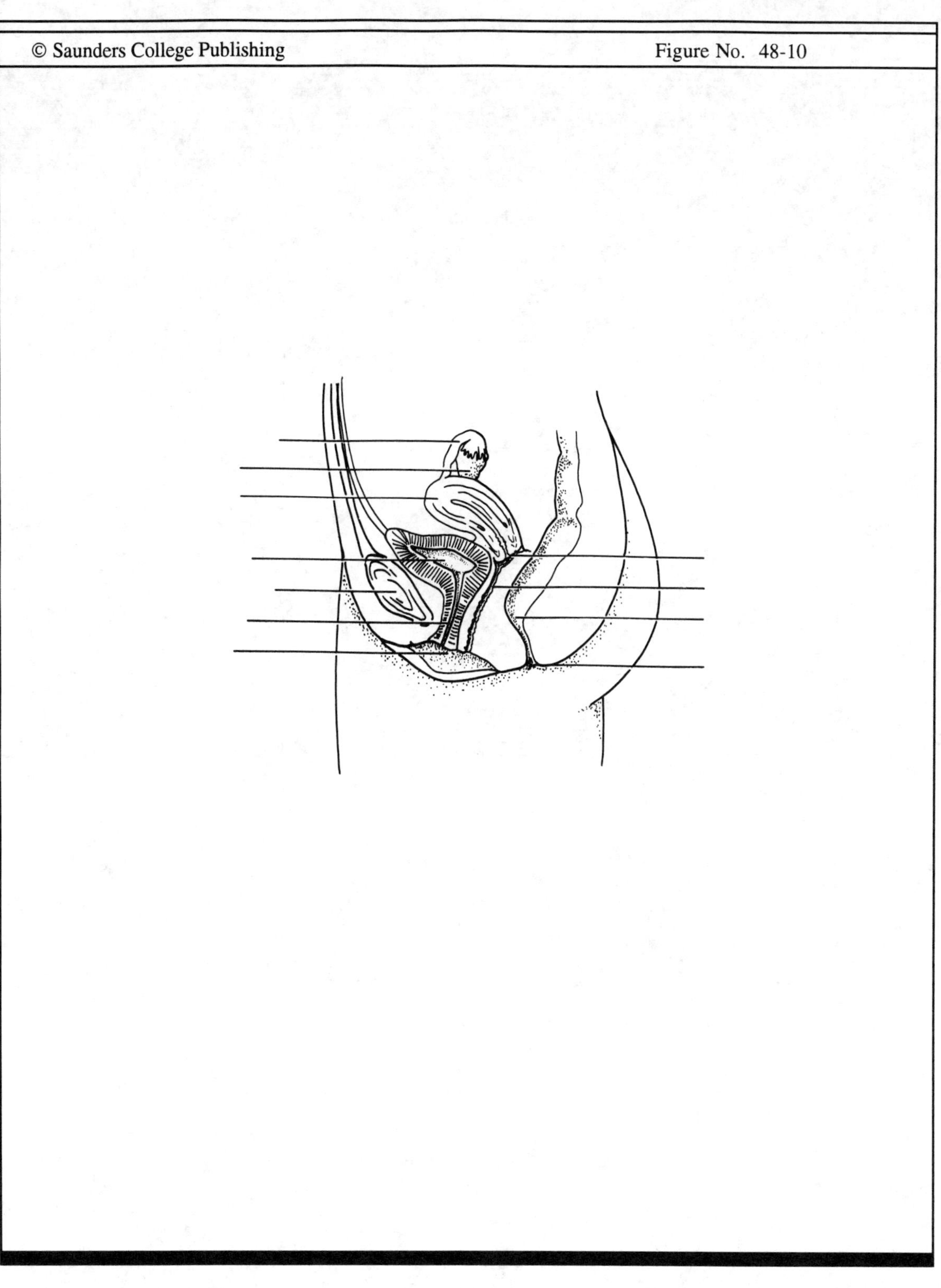

Figure No. 48-11

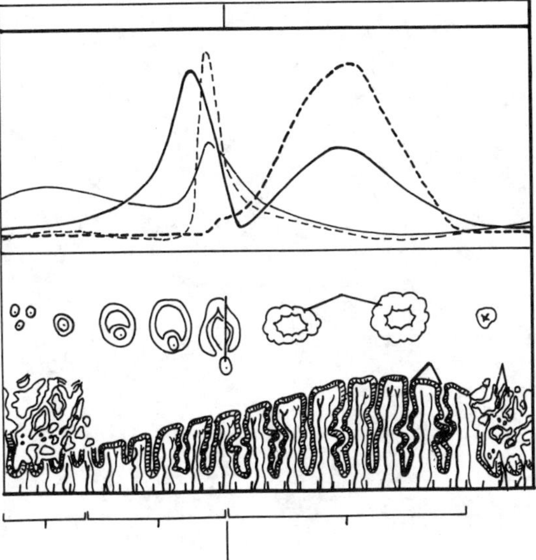

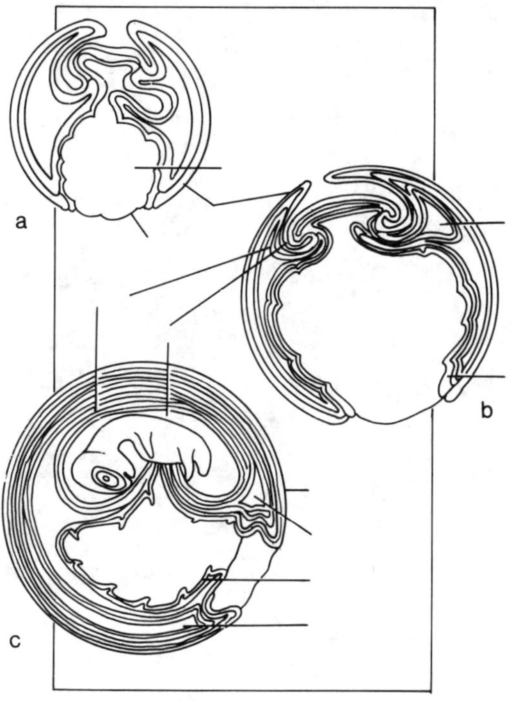

a
b
c

   Figure No. 54-1

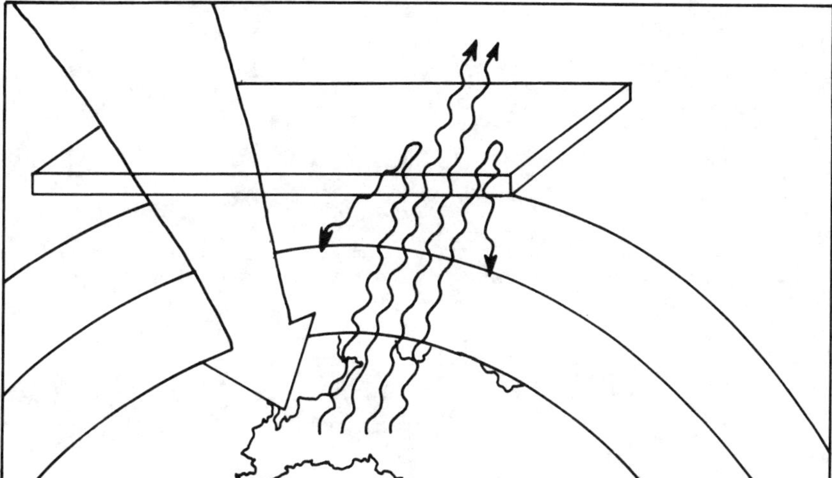

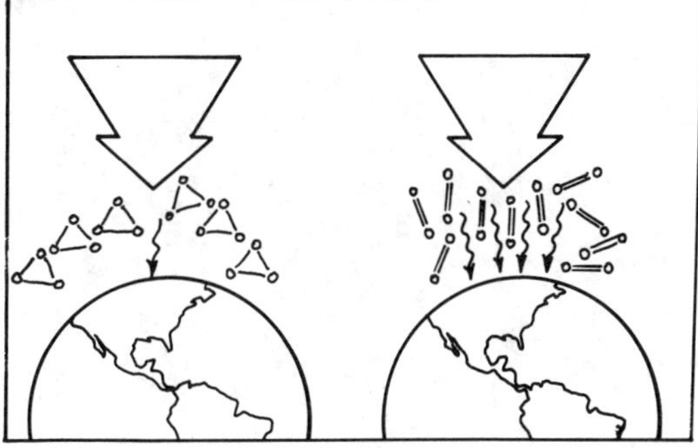